Mass Transfer in Chemical Engineering Processes

Mass Transfer in Chemical Engineering Processes

Editor

Hemming Lindberg

scitus academics

Mass Transfer in Chemical Engineering Processes
Edited by **Hemming Lindberg**

ISBN: 978-1-68117-270-5
Library of Congress Control Number: 2016937440

SCITUS Academics LLC,
www.scitusacademics.com
Box No. 4766, 616 Corporate Way,
Suite 2, Valley Cottage,
NY 10989

Notice

Reasonable efforts have been made to publish reliable data and views articulated in the chapters are those of the individual contributors, and not necessarily those of the editors or publishers. Editors or publishers are not responsible for the accuracy of the information in the published chapters or consequences of their use. The publisher believes no responsibility for any damage or grievance to the persons or property arising out of the use of any materials, instructions, methods or thoughts in the book. The editors and the publisher have attempted to trace the copyright holders of all material reproduced in this publication and apologize to copyright holders if permission has not been obtained. If any copyright holder has not been acknowledged, please write to us so we may rectify.

Printed in United States of America on Acid Free Paper

Preface

Mass transfer describes the net mass movement from one location to another, which usually means stream, phase, fraction or component. In many processes such as absorption, evaporation, adsorption, drying, precipitation, membrane filtration and distillation, mass transfer occurs. Mass transfer is used for various processes and mechanisms by various scientific disciplines. The phrase is commonly used in engineering for physical processes involving the convective and diffuse transport of chemical species in physical systems. The theory of mass transfer makes it possible to calculate mass flow in a system and to distribute the mass of different species over time and space in such a system, even when chemical reactions occur. The purpose of these calculations is to understand such a system and possibly to design or control it. The evaporation of water from the lake into the atmosphere, the purification of blood in the kidneys and liver and the distillation of alcohol are some common phenomena of mass transfer procedures. Mass transfer operations involve the separation of chemical components in distillation columns in industrial processes. In additional transport processes, such as in industrial cooling towers, mass transfer is often linked to. These towers combine heat transfer with hot water to flow with hotter air and evaporate as heat flows from the air.

This book, Mass Transfer in Chemical Engineering Processes, compromises several approaches to solving problems of mass transfer for various practical applications in chemical engineering. For its readers, the book should be of great importance with interesting ideas and inspirations or direct solutions to their specific problems.

Table of Contents

Chapter 1

Mass Transfer in Infrared Drying of Gel-Coated Seeds

Antônio M. Barbosa Neto[1], Luanda G. Marques[1], Manoel M. Prado[1*], Dermeval J. M. Sartori[2]

[1]Department of Chemical Engineering, Federal University of Sergipe, São Cristóvão, Brazil

[2]Department of Chemical Engineering, Federal University of São Carlos, São Carlos, Brazil

ABSTRACT

In order to contribute for a better understanding of mass transfer in drying of shrinking particles, in this study shrinkage and drying characteristics of sorghum seeds encapsulated into gel-based polymeric matrix were experimentally determined by using infrared (IR) radiation. IR drying of gel coated seeds was carried out at three different temperatures (65°C, 80°C and 93°C). The shrinkage of the individual particles during drying was quantified by means of the volume and surface area changes evaluated from geometric measurements. The product quality was evaluated in terms of the changes of particle density and percent of cracks in gel coating incurred during drying.

Surface area and volume of the gel-seed system decreased about 65% and 80% until the end of the process, respectively, stressing the need to take into account the surface area changes to calculate water flux density as function of moisture content and obtain an accurate interpretation of the drying mechanisms well as to include the volume shrinkage in mass transfer models to determine reliable values of moisture diffusivity. The IR drying behavior of gel-coated seeds was then characterized by the presence of three drying periods: heating up, constant moisture flux and falling moisture flux. Accelerated drying of gel coated seeds was obtained by applying higher IR radiation intensities. The effect of IR source temperature on the particle shrinkage was more pronounced at the constant moisture flux period and practically negligible at the decreasing moisture flux period. Neglecting shrinkage of individual coated-seeds during IR drying led to an erroneous absence of constant flux period and overestimation of the mass transfer by diffusion. Apparent density of the particles was greater at low temperature IR drying than at high-temperature IR drying. Coated particles keep their original geometry, but a significant cracking of gel coating was observed at rapid drying rate conditions.

KEYWORDS

Gel Coating; Shrinking Particles; Infrared Drying; Mass Transfer

1. INTRODUCTION

Sorghum [Sorghum bicolor (L.) Moench] is one of the main agricultural crops in the world scenery. It is cultivated in Brazil predominantly for the production of grain and forage, but a growing interest is being focused on the potential of sorghum to produce bioethanol feedstocks [1]. Unlike sugar cane, sorghum is grown from seed and has a much shorter growing season of around 120 to 130 days. It is also capable of growing on marginal lands where other crops do not thrive.

Breeding programs for new varieties of sorghum suited to semi arid tropics are under development. However, in the semiarid regions of the Brazilian Northeast, water and salinity stresses are primary limiting environmental conditions which restrict successful establishment of crops. Seed germination may be significantly reduced and delayed by the decreasing rate of water absorption, when drought stress and soil salinity are high or when saline water is used for irrigation [2].

Among the numerous seed treatment methods, the coating with hydrocolloids due to their hydrophilic characteristics can be viewed as a promising technique to be performed on sorghum seeds, in order to overcome constraints to seed germination, as the water stress in soil, a common problem in areas of Brazilian Northeast, with low rainfall [3]. Polymer-based coating may also reduce the leaching of inhibitors from the seed coverings as well as restrict salts and oxygen diffusion to the embryo [4]. In addition, polymer coating can regulate moisture sorption during storage, providing protection against insect infestation and microbial attack and improving seed conservation. Hydrocolloid-based polymeric matrices such as agar and alginate gels have been successfully employed for encapsulating seeds of many crops, improving their germination and emergence [5,6].

Drying is an essential operation in seed coating since this step ensures moisture removal from the coating, promoting the gel adherence to the seed surface and preventing deterioration due to molds infestation. According to Kajtna et al. [7], the drying process does not improve the adhesion properties of the coating, which are dependent on the polymer properties. However, improperly chosen drying conditions may cause polymer degradation and worsen the adhesion properties, leading to losses in coated seed quality.

The increased awareness about environmental degradation has motivated the application of alternative sources of clean energy in drying operations [8]. Within this context, drying using infrared (IR) radiation appears as an interesting method to intensify heat and mass transfer rates, to minimize energy consumption and to preserve quality of coated seeds. The layer-layer drying of the material by applying IR radiation can prevent transfer of soluble substances beyond the product surface and formation of surface films, and the relatively high rate of drying suppresses oxidation and prevent heatsensitive substances being lost from the product during drying. Preserving the integrity of the gel coating structure allows maintaining its ability to imbibe water, which is an essential quality attribute of coated seeds in the germination process.

The application of IR radiation for drying purposes in chemical and agro-food industries has become widely widespread, due to its intrinsically useful characteristics such as higher energy efficiency, shorter drying time, versatility, simplicity of equipment, fast response of heating and drying, easy incorporation to any existing dryer and low capital cost [9]. A lot of work has been devoted to the research on IR drying of various materials, such as: paper [10], polymers [7], pharmaceutical films [11], paints and pigments [12,13], hydrous ferrous sulphate [14], and agricultural products [15-18]. However, there is still no information available in the literature regarding the use of IR radiation for drying systems composed by seeds coated by gel-based polymeric matrices.

The infrared drying of gel-coated particles is a challenging assignment due to the complex relationship between the transport phenomena and the shrinkage characteristics, which are directly related to quality attributes, such as density, porosity, sorption characteristics, crust and cracks [19]. Due to the highly deformable polymeric structure of gel coating, gel coated seeds undergo changes in their original dimensions and shape during moisture removal, modifying particle structural properties as well as the heat and mass exchange area, thus affecting the drying rate. The complexity increases as the extent of shrinkage is also process dependent. That is the result of the moisture gradient in the product, which, in turn, induces stresses and, thus, mechanical deformation [19].

Thus, any attempt to raise the shrinkage phenomenon to a higher level of understanding must address the structural properties of the material such as density, porosity, specific area and surface cracking.

The changes in shape, dimensions and solid structure of deformable particles yield to a particular system where the available coefficients of mass transfer are not suitable to reproduce moisture loss during drying of the material. Thus, any attempt to accurately interpret the mass transfer phenomenon during drying of shrinking particles like gel coated seeds requires an experimental investigation to obtain reliable values of effective moisture diffusivity (D_{eff}), which should take into account the reduction in the distance required for the movement of water molecules. However, few works have been carried out on the effects of the shrinkage phenomenon on D_{eff}, even less on $D_{eff}$of gel coated seeds.

The quantification of the particle shrinkage and its dependence on process variables is essential to improve the understanding of drying kinetics and hence to provide more realistic information on the mechanism of moisture transport, the influence of operating conditions on the drying behavior as well as for selection of optimal drying conditions for seed quality control [20,21].

The objectives of this work were, therefore, to determine the shrinkage and drying characteristics of agar gelcoated sorghum seeds subjected to infrared radiation as well to evaluate the influence of particle shrinkage on the effective moisture diffusivity. The effects of IR radiation intensity on the percent of cracks in gel coating were also examined.

2. MATERIALS AND METHODS

2.1. Materials

In this study, sorghum (Sorghum bicolor L. Moenche) seeds from Embrapa BRS-330 variety were used. This early duration variety is tannin-free and is characterized by high protein content. It was grown at Experimental Station of Embrapa Maize and Sorghum, Sete Lagoas/ MG.

The materials used for forming coating matrices included agar agar (Acumedia Manufactures, Maryland, USA) as gelling agent, sorbitol (Merck) as plasticizing agent, starch powder (Vetec Química Fina Ltda) as binder, and water as solvent. Filler agents of the coating, as cellulose, talc and titanium dioxid, were generously provided by Minérios Ouro Branco Ltda, São Paulo, Brazil.

2.2. Preparation of Materials

Sorghum seeds were received from Embrapa Maize and sorghum with a moisture content of 0.126 wet basis (d.b.), at room temperature, and free from broken grains and foreign material. Seed size distribution was firstly determined from sieve analysis. Medium seeds (sieve diameter between 3.6 and 4 mm) were found to represent about 80% population in the lot. Thus they were chosen for the study, in order to ensure uniform particle size.

The coating mixture was prepared in accordance with the Committee on Food Chemicals Codex [22]. It had the following formulation: 1.5% agar, 0.5% sorbitol, 1.5% starch, 4% de cellulose, 0.5% titanium dioxid and 92% water. This formulation provided the best adherence of the coating mixture to the seed surface.

2.3. Coating Methodology

Application of the polymeric mixtures in rectangular moulds containing the seeds was the adopted technique for entrapping the sorghum seeds within gel matrices. The beaker containing the polymeric mixture was immersed in a thermal bath at 50°C to maintain the material in gum phase. After poured into the moulds the coating mixtures were allowed to gel at room temperature for 2 h to enable high gel strength prior to removing gel-seed system from the moulds. This resulted in the formation of encapsulated seeds, rectangular in shape 6 mm × 12 mm × 18 mm (thickness × width × length).

2.4. Drying of Gel-Seed System

The experimental set-up developed to study the single layer drying of gel coated sorghum seeds using infrared radiation is shown schematically in Figure 1. The experimental unit consists of a drying chamber equipped with a 250 W incandescent lamp. The drying chamber of 200 × 200 × 300 mm was made from wood sheet of 10 mm thickness. The chamber walls were isolated and covered with an aluminum foil. The temperature of the IR source was regulated using a temperature controller. The distance between the infrared heating source and the monolayer of particles was set at 17 cm. Drying tests were conducted with IR source temperatures of 65˚C, 80˚C and 93˚C.

Two experiments groups were carried out. The aim of the first group was to determine the drying kinetics of the gel-coated seeds at different IR radiation intensities in terms of the moisture content and temperature evolution.

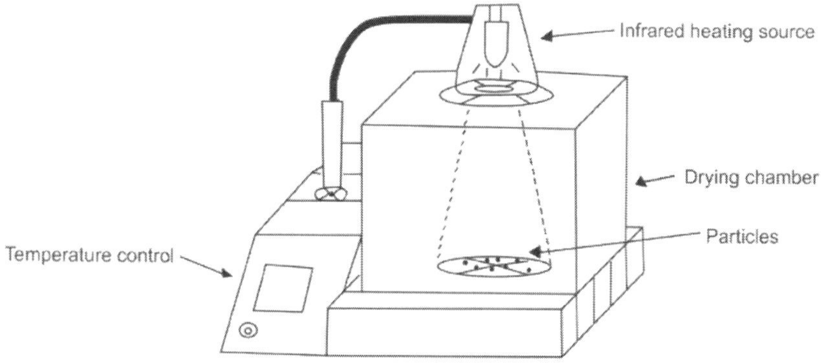

Figure 1. Schematic diagram of the experimental set-up.

The second group had as objective to identify and quantify the shrinkage of the gel-coated particles as a function of moisture content and IR drying conditions.

The dryer was run empty for about 0.5 h to attain a steady state in respect of the pre-set operating conditions before each drying test. About 22 g of sample were used at each experiment. Seeds were uniformly spread on the aluminum tray (130 mm diameter) in a mono layer. Each experiment and subsequent analyses were performed in triplicate.

To determine the drying kinetics curves the seed samples were removed from the equipment at regular intervals along the total drying time and their weight measured in a digital balance (Sartorius, CP224S, Germany) with 10^{-4} g accuracy. All weighting processes were completed in 15 s during the drying process. Moisture content at a given time was calculated by the ratio between the mass of evaporated water and the dry mass, which was determined at the end of each experiment by the oven method at 105°C ± 3°C for 24 hours. The surface temperature of the particles was measured at different drying times by using an optical pyrometer.

Coated seeds samples were subjected to IR radiation for different periods to achieve the desired moisture content, with previously drying kinetics curves facilitating determination of the drying time. On attaining the predetermined drying time, the coated seeds were taken from the dryer and their dimensions were measured using a digital micrometer (Pantec, Model 13101-25, accuracy of 10^{-3} mm). The formation of cracks in the coating structure during drying was quantified by image analysis. Percent of breakage was then used as quality attribute of the coated seeds.

2.5. Analysis of Drying Kinetics and Particle Shrinkage Data

The experimental data sets from the different drying tests were initially expressed as moisture content (X) versus time (t). The moisture contents values were also converted into the dimensionless moisture ratios (XR), defined as the ratio of the free water still to be removed at any time to the total free water initially available, and expressed as:

$$XR = \frac{\bar{X} - X_e}{X_0 - X_e} \quad (1)$$

where $\bar{X}$ is the average moisture content at a given time, X_0 and X_e are the initial and equilibrium moisture content of coated particles, respectively. All at dry basis. For IR drying data the value of X_e in Equation (1) was set to zero since prolonged exposure of agricultural material to infrared radiation will eventually cause a complete moisture removal [23].

The shrinkage of gel coated seeds during single layer drying was quantified from the changes in volume and surface area of individual particles, which were calculated from the dimensions taken by the digital micrometer. Reduced dimensional changes in both volume (V_p/V_{p0}) and area (A_p/A_{p0}) with respect to the moisture content constituted the so-called shrinkage curves. It must be pointed out that the shrinkage analysis, based on volume and surface area changes from the geometric measurements with the aid of a digital micrometer, was sufficiently accurate since gel coated seeds keep its original shape throughout the whole drying process.

The water flux density, averaged over the entire exchange surface, was then calculated at a given time or moisture content using the following equation:

$$N_w = -\frac{m_{ds}}{A_p}\frac{\mathrm{d}\bar{X}}{\mathrm{d}t} \tag{2}$$

where N_w is the mass flux density (kg·m^{-2}·min^{-1}), m_{ds} is the weight of the dry solids (kg), and $A_p$is the surface area of the particles exposed to IR radiation (m^2). The drying rate per mass unit of dry solid ($-\mathrm{d}\bar{X}/\mathrm{d}t$) was determined from the drying curves by numerical derivation.

The mass transfer in IR drying of gel-coated seeds was analyzed in terms of effective diffusivity, which was determined at each IR source temperature by applying diffusive model to describe the drying kinetics under the consideration of both negligible and significant shrinkage.

Non-steady mass transfer equation based on Fick's second law for the diffusion of water during drying is expressed as

$$\frac{\partial(\rho_d X)}{\partial t} = \nabla\cdot\left[D_{ef}\nabla(\rho_d X)\right] \tag{3}$$

where D_{ef} is the effective diffusivity, r_d is the local concentration of dry solid (kg dry solid per volume of the moist material) that varies with moisture content, because of shrinkage and X is the local moisture content (dry basis).

Assuming one-dimensional moisture transfer with negligible shrinkage and considering that the process is isothermal, initial moisture distribution is uniform, effective water diffusivity into material is constant, and equilibrium conditions at the particle surface, the analytical solution for rectangular geometry, if only the first term is considered, can be approximated to the form [24]:

as

$$XR = \frac{\bar{X} - X_e}{X_0 - X_e} = \frac{8}{\pi^2}\exp\left(-\frac{\pi^2 D_{eff}}{L_0^2}t\right) \tag{4}$$

where L_0 is the slab thickness and D_{eff} is the effective moisture diffusivity without considering the particle shrinkage.

In order to include the shrinkage effects, substituting the density of dry solid ($\rho_d = m_d/V_p$), Equation (3) for constant mass of dry solid could be expressed as [20,25, 26]:

$$\frac{\partial\left(X/V_p\right)}{\partial t} = \nabla\cdot\left[D_{ef}\nabla\left(X/V_p\right)\right] \tag{5}$$

where V_p is the particle volume. Substituting $Y = X/V_p$, the following equation is obtained:

$$\frac{\partial Y}{\partial t} = \nabla\cdot\left(D_{eff}^{*}\nabla Y\right) \tag{6}$$

For one-dimension moisture transfer in conjunction with the following initial and boundary conditions

$$t = 0,\ Y = X_0/V_0 \tag{7}$$

$$t > 0,\ z = 0,\ \frac{\partial Y}{\partial z} = 0 \tag{8}$$

$$t > 0,\ z = L,\ Y = X_{eq}/V_{eq} \tag{9}$$

a solution similar to Equation (11) is obtained:

$$YR = \frac{\bar{Y} - Y_e}{Y_0 - Y_e} = \frac{X/V - X_e/V_e}{X_0/V_0 - X_e/V_e} = \frac{6}{\pi^2}\exp\left(-\frac{\pi^2 D_{eff}^*}{L^2}t\right) \tag{10}$$

where V_p is the sample volume at a given time, V_{p0} is the initial volume and V_{pe} is the sample volume on attaining the equilibrium moisture, D_{eff}^* is the effective moisture diffusivity of shrinking particle and L is the time averaged thickness during drying.

Equations (4) and (10) are valid only for describing the falling rate period when the drying process is controlled by internal moisture diffusion and moisture content is below the critical value. Therefore, the water diffusivity into gel-coated seeds undergoing IR drying was estimated from these equations by setting the initial moisture content to the critical value X_{cr} and by setting the drying time to zero when the mean moisture content of the sample reaches that critical moisture content. Experimental data of XR and YR were then fitted to the Equations (4) and (10), respectively. D_{eff} and D_{eff}^* -values were estimated by applying a non-linear regression procedure using STATISTICA® software. The estimation method was based on the well-established Levenberg-Marquardt algorithm [27]. The statistical criteria used to evaluate the goodness-of-fit of Equations (4) and (10) were the coefficient of determination (R^2) and the root mean square error (RMSE). RMSE was calculated from the predicted and the experimental XR values as follows:

$$\text{RMSE} = \sqrt{\frac{1}{n}\sum_{i=1}^{n}\left(XR_{pred,i} - XR_{exp,i}\right)^2} \tag{11}$$

where n is the number of data points.

The dependency of effective diffusivity on temperature was investigated by plotting the obtained values against product temperature. The trends were visually inspected and appropriate model was fitted to the data.

3. RESULTS AND DISCUSSION

3.1. Shrinkage Characteristics

Experimental shrinkage data of gel coated seeds, expressed by A_p/A_{p0} and V_p/V_{p0} ratios, were plotted against the dimensionless moisture content at different IR radiation source temperatures and are presented in Figures 2 and 3, respectively.

A steep rise in area and volume reduction of coated seeds samples was observed at the early and intermediate drying stages. On attaining a critical XR-value (around 0.2), there is a decrease in slope of shrinkage curves and both the area and volume contraction in particles increases slowly. This change in shrinkage behavior is due to modifications in the proportion of external and internal resistances to moisture transfer [28], which are caused by changes in the mechanism of mass transfer, what will be discussed in the following section. In the final stage of drying internal resistances control the mass transport and the migration of internal moisture is not uniform, so that it does not compensate the evaporation rate from the particle surface. Formation of a rigid crust in the gel coating may then be induced, limiting the particle volume, thus hindering subsequent shrinkage. This may also be attributed to the decrease at low moisture contents in the mobility of the polymeric matrix coating the seeds [29].

Particle shrinkage was found to increase with decrease in temperature of IR source, but this effect was observed to be more pronounced at the intermediate stage, until XR has reached the critical value. For example, at a XR-value of 0.37, the reductions in both particle area (Figure 2) and volume (Figure 3) at 65°C were nearly 23% and 25% higher than that for the seeds dried at 93°C. Exposing the particles under low drying temperatures results in low drying rates, so that external resistance controls the mass transfer, moisture profiles into the particles are uniform and stresses are minimum [28]. This contributes for the shrinkage phenomenon.

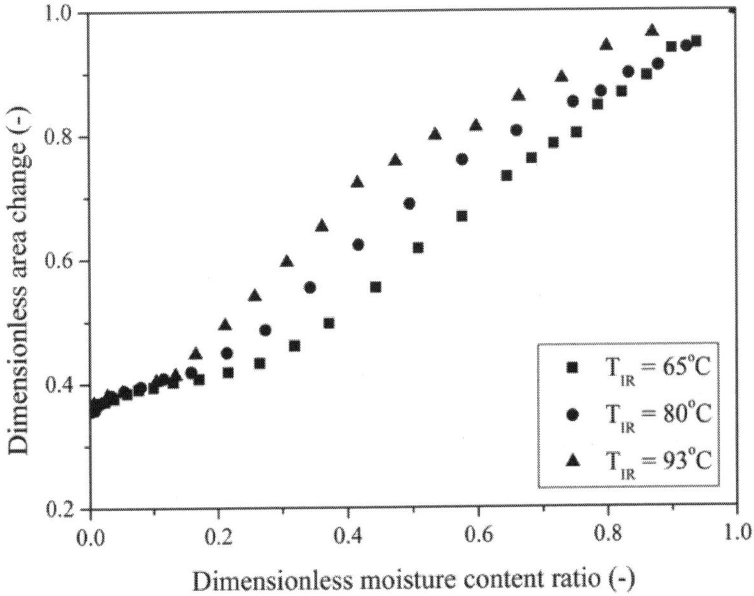

Figure 2. Dimensionless area change (A_p/A_{p0}) of gel-coated seeds as a function of dimensionless moisture content (XR) during IR drying at different temperatures.

Below the XR of about 0.2, the deviation within the experimental data (Figures 2 and 3) obtained under different IR-drying temperatures was smaller than the measurement uncertainties. This indicates that for the final stage of drying the temperature had a negligible effect on the shrinkage of gel coated seeds, so that both volume and surface area changes can be correlated with only the averaged moisture content of the dehydrated seeds.

At the end of IR-drying the reductions in volume and surface area of gel coated seeds for the investigated temperature range were high, approximately 80% ± 2% and 64% ± 1%, respectively, stressing the need to take into account the surface area changes to the calculation of the water flux density as well as to include the shrinkage in the mass transfer models.

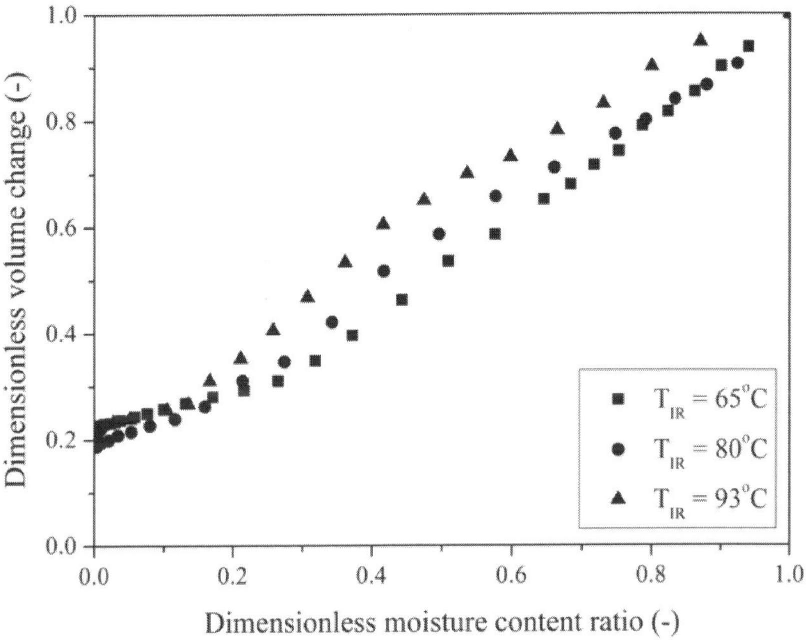

Figure 3. Dimensionless volume change (V_p/V_{p0}) of gelcoated seeds as a function of dimensionless moisture content (XR) during IR drying at different temperatures.

3.2. Drying Characteristics

Typical results of surface temperature and water flux density as a function of time, calculated with and without consideration of reduction in particle area exposed to IR radiation are shown in Figure 4.

Because of the high initial water content of gel-coated seeds as well as of the deformable coating structure, the course of IR drying was accompanied by both shrinking surface area and volume as discussed previously. As a result, there is a large area to volume ratio for heat and mass transfer, facilitating the moisture transport, thus leading to higher values of water flux density. On the other hand, if the reduction in mass exchange area is neglected lower values for moisture flux density are obtained and it decreases even before the completion of the first stage of drying, although free water remains available on the surface [30]. It is evident from Figure 4 that on considering constant mass transfer area no constantflux period was observed, providing a misleading interpretation of the drying mechanisms.

It should be pointed out that the presence of a constant flux period has been rarely noticed in the most of studies on IR drying [7,9,14,17], probably because mass exchange area during drying are generally not considered [31].

The analysis of drying kinetics in terms of the evolution in surface particle temperature in conjunction with the water flux density indicates that the drying behavior of gel-coated seeds exposed to IR radiation is characterized by the presence of three periods of drying: heating up, constant flux and one falling flux period.

At the beginning of drying process, under all the operating conditions, there was a warming-up period, during which both the water flux density and the temperature increase. This initial stage represents a transition period corresponding to non-isothermal conditions. The heating period was followed by a constant flux period, in which the heat and mass flux densities are equal. During this period migration of internal moisture compensates the evaporation rate from the particle surface, maintaining this surface saturated with a moisture film. As a consequence, all infrared energy is used as latent heat to vaporize the surface moisture in gel-seed system and drying occurs under a constant saturation temperature, as evidenced by the temperature plateau in Figure 4. From the critical moisture critical, the aforementioned moisture compensation ceases and part of the radiant energy is used in the form of sensible heat, causing an increase in solid temperature, while the moisture flux density decreases with reduction in the moisture content, thus characterizing the falling flux period, which is mainly controlled by moisture diffusion within the particles. This is a complex mechanism involving water in both liquid and vapor states, which is commonly characterized by a socalled effective diffusivity [24].

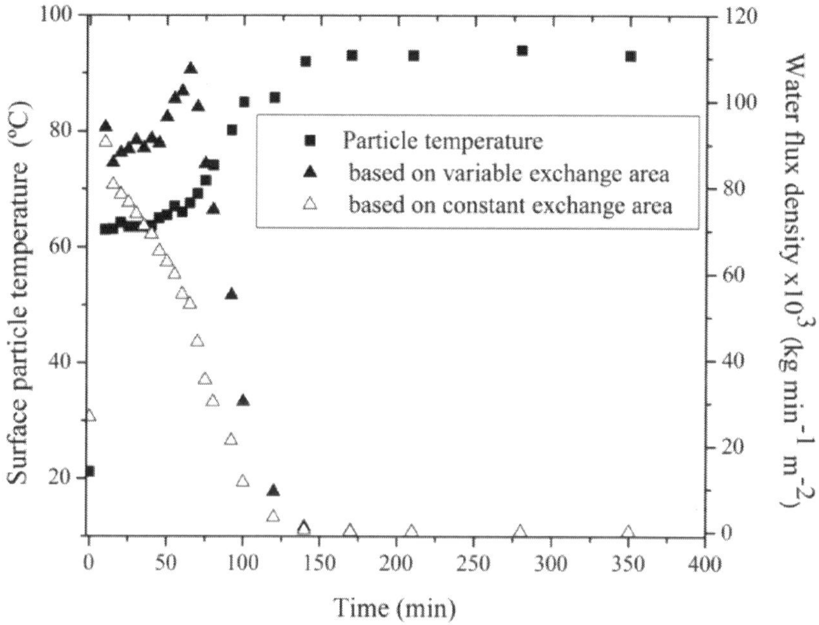

Figure 4. Surface temperature and water flux density of coated particles as a function of time during IR drying at 93°C.

The changes in drying behavior presented in Figure 4 also provide evidences on the reduction in IR absorption capability as the moisture of the seeds was removed. High initial moisture content of the gel-coated seeds contributed for higher absorption of radiation energy at the early stage of drying, resulting in a raised heat content of water inside the seed. The water tends to vigorously flash out by evaporation, thus taking the latent heat of vaporization from the seed, consequently increasing the drying rate [17]. During the constant rate period, when internal moisture migration compensates the evaporation of moisture from the seed surface, mass and heat transfer rates are equal. At the falling rate period, with the particle surface drying out, heat penetration through the dried layer is decreased, thus reducing the drying rate [32].

Typical results of moisture flux density versus moisture content for gel-coated seeds samples dried using infrared radiation at different heating source temperatures are shown in Figure 5. As expected, an increase in IR source temperature resulted in an enhancement in the overall water flux density and thus a reduced total drying time. This can be attributed to a rapid increase in particle temperature, caused by a higher intensity of radiant energy applied, resulting into an increase in the water vapor pressure inside the product and thus in higher drying flux densities [33].

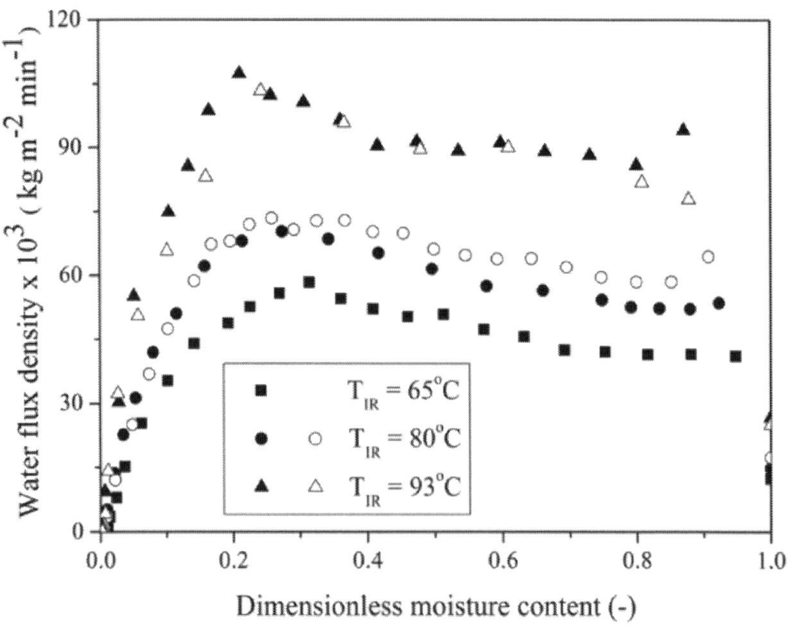

Figure 5. Moisture flux density versus moisture content during IR drying of gel coated seeds at different IR source temperatures.

An increase in drying rates with an increase in IR radiation power or intensity has commonly been reported in the literature [7-18].

Table 1 summarizes the values of moisture flux density during the constant flux period ($N_{w,c}$), critical moisture content (X_{cr}), and critical drying time (t_{cr}) for all IR source temperatures (T_{IR}) investigated in this study.

As expected, an increase in IR source temperature resulted in a reduction of the constant flux period. However, the moisture content at which the transition between constant and falling flux periods occurred was found to be practically independent on drying conditions.

In addition, based on the critical values of XR reported, it can be stated from Figures 2 and 3 that shrinkage of particles has occurred mainly in the constant flux period.

The constant flux period was then represented by a coefficient of mass transfer (k_c), whose relationship with the IR source temperature is expressed as:

$$N_{w,c} = k_c = 2.14\times10^{-5}T^{1.859} \tag{12}$$

In what follows, the analysis of the decreasing flux period based on the application of the diffusive model is presented, aiming to evaluate the influence of particle shrinkage on the estimation of effective moisture diffusivity.

Table 1. Summary of t_{cr}, X_{cr} and $N_{w,c}$ data for single layer IR drying of gel-coated sorghum seeds.

T_{IR} (°C)	t_{cr} (min)	X_{cr} (d.b.)	XR_{cr} (-)	$N_{w,c}\times10^3$ ($kg\cdot m^{-2}\cdot min^{-1}$)
65	145 ± 5	1.477 ± 0.117	0.209 ± 0.016	52 ± 3
80	95 ± 5	1.394 ± 0.123	0.205 ± 0.009	71 ± 1
93	62 ± 3	1.539 ± 0.162	0.211 ± 0.006	99 ± 4

3.3. Influence of Shrinkage on Effective Moisture Diffusivity

Diffusive models without and with shrinkage, Equations 4 and 10, were then used to estimate the effective diffusion coefficient of gel-coated seeds subjected to IR-drying. Table 2 shows the results obtained for the three IR source temperatures investigated in this study.

It can be verified that the values of diffusivity calculated without consideration of the shrinkage were higher than those obtained taking into account the phenomenon. For example, at 65°C the effective diffusivity of coated seeds without considering the shrinkage (D_{eff} = 12.1 × 10^{-11} m^2/min) was found to be 116% higher than that estimated taking into account the phenomenon (D^*_{eff} = 5.51 × 10^{-11} m^2/min).

Neglecting shrinkage of coated seeds during IR drying leads, therefore, to an overestimation of the mass transfer by diffusion. The explanation is on the fact that particle shrinkage produces a variation in the distance required for the movement of water molecules, therefore making the effective diffusivity be overestimated when obtained from analytical solution of diffusion model without considering volume contraction. This finding agrees with results of previous reports on other products [25,26].

The temperature dependency of the effective diffusivities estimated considering shrinkage was described by an Arrhenius-type relationship. The activation energy for infrared-dried coated seeds was found to be 33 kJ/mol, which is within the range of 12 - 110 kJ/mol reported for several agricultural materials [15-18]. The Arrhenius factor was 5.161 × 10^{-5} m^2/s, with R^2 of 0.9077.

3.4. Changes in Structural Properties

The shrinkage-induced changes in structural properties of the gel-coated seeds were analyzed in terms of their apparent density. Typical results on particle density as a function of moisture content are shown in Figure 6.

An explanation for the change in behavior of particle density during IR-drying may be ascribed to the distinct characteristics of reduction in mass and volume of the coated seed during moisture removal, which were previously discussed.

In the whole range of temperatures investigated, coated seeds had its apparent density almost constant during the constant flux period, due to a reduction of both particle volume and amount of water evaporated in the same proportions. In the decreasing flux period as the volume contraction of particle is limited particle density steadily decreased with the loss of moisture. This result is an indicative that the pore formation within the gelbased coating structure begins at moisture contents below the critical moisture content. Similar behavior has been reported during IR drying of quince [34].

It can also be seen from Figure 6 that the IR drying temperature affected the apparent density of coated seeds. Apparent density of gel-coated seeds produced after IR drying at 65°C was about 17% higher than that obtained at 93°C. The denser structure obtained at 65°C suggests that there was a reduction of pore volume within the gel based coating structure as a result of a more uniform shrinkage during slow drying rate conditions. On the other hand, under rapid drying rate conditions, particle surface becomes fast unsaturated in moisture, so that low moisture content of this external surface may induce a rubber-glass transition and the formation of a porous outer rigid crust or shell that fixes the volume and hinders subsequent shrinkage of the still rubbery, inner part of the gel coating [29]. However, this contributes for replacement of evaporated water by air, thus leading to lower values of particle density.

Table 2. Values of effective moisture diffusivity estimated without and with consideration of shrinkage during IR drying of sorghum seeds encapsulated into agar gel.

T_{IR} (°C)	Without shrinkage			With shrinkage		
	$D_{eff} \times 10^{11}$ (m^2/s)	R^2	RMSE	$D^*_{eff} \times 10^{11}$ (m^2/s)	R^2	RMSE
65	7.57	0.9959	0.0066	4.55	0.9943	0.0258
80	12.1	0.9964	0.0051	5.51	0.9899	0.0362
93	20.7	0.9953	0.0050	11.3	0.9828	0.0471

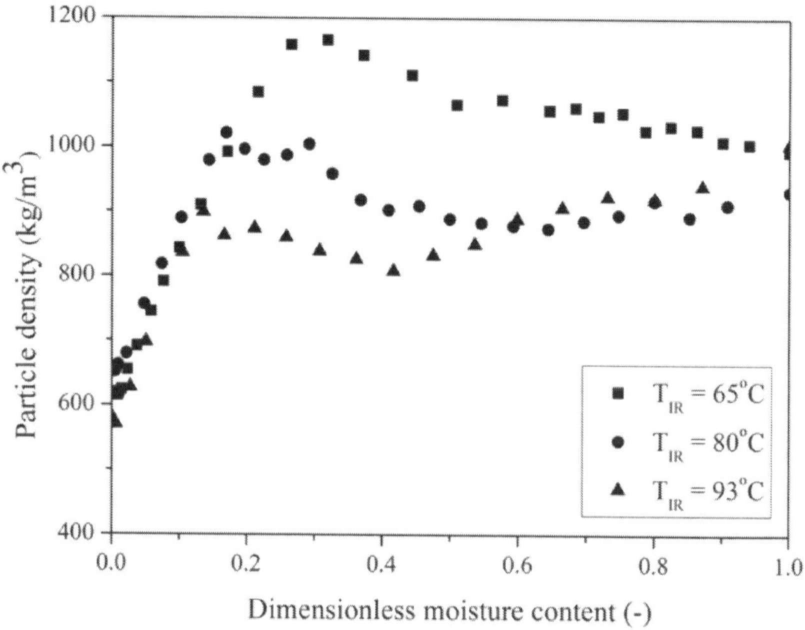

Figure 6. Particle density as a function of moisture content, for different IR source temperatures.

The quality of the coated seeds produced at each temperature was evaluated in terms of the breakage percent in the artificial coating. Figure 7 shows the percent of breakage in coated particles versus the temperature employed in the drying process.

The percent of breakage ranged from 8% to 42% as the drying temperature increased from 65˚C to 93˚C. This increase in cracking might be attributed to the marked increase in water flux density as the IR heating temperature increased (Figure 5). Rapid drying rate conditions could have contributed to intense moisture gradients within the gel-based coating and, thus, to a higher state of stress in the gel-seed system, resulting in a high percent of breakage.

To produce gel-coated sorghum seeds with their noncracking structures mild drying conditions must therefore be applied. Modifying the composition of the coating matrix is another alternative, in order to retard its loss of elasticity during drying, thus increasing the resistance to breakage.

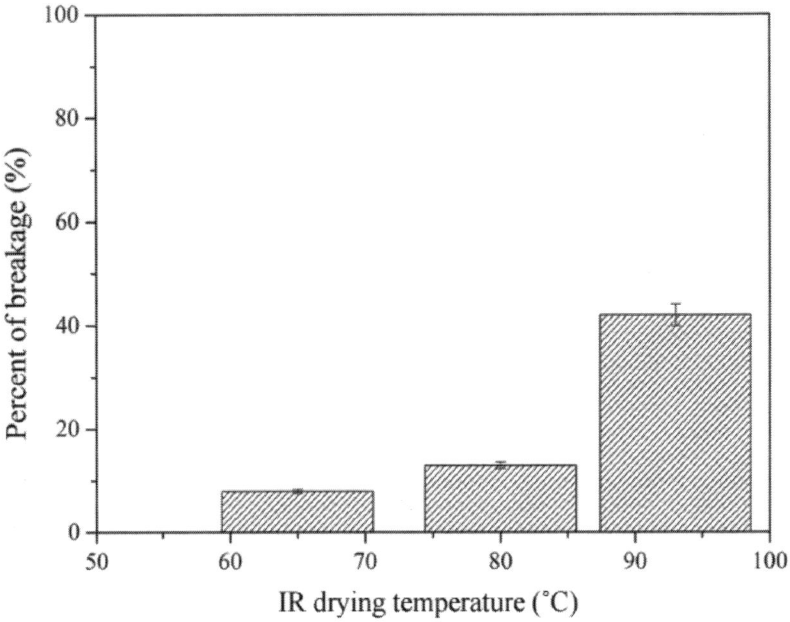

Figure 7. Percent of breakage in coated seeds during IR drying at different temperatures.

4. CONCLUSIONS

The analysis of drying kinetics based on both evolutions of solid temperature and water flux density taking into account the changes in mass exchange area is essential to obtain an accurate interpretation of mass transfer in IRdrying of shrinking particles like sorghum seeds encapsulated into gel-based polymeric matrix.

The IR drying behavior of gel-coated seeds was characterized by the presence of three periods of drying: warming-up, constant flux and one falling flux period.

Gel-coated seeds subjected to drying using infrared radiation had significant shrinkage, which occurred mainly during the constant flux period. In this stage of drying, the particle density was almost constant with reduction in moisture content. The decrease in apparent density during the falling moisture flux period evidenced a volume contraction of particles lower than the amount of removed water.

Changes in surface area and volume of coated seeds were found to be affected by IR source temperature during the constant flux period, when an increase in temperature reduced particle shrinkage. At the final stage of drying, the aforementioned changes were dependent only on the average moisture content of particles.

Neglecting the shrinkage of the particles led to a significant overestimation of the effective moisture diffusivity and thus of the energy required for drying the material.

IR drying temperature significantly affected apparent density of gel-coated seeds and this influence is closely related to the rates at which both the volume shrinkage and drying occur. Particles dried under rapid drying rate conditions had the lowest apparent density values.

Percent of cracks in gel-coated seeds produced by high-temperature IR drying was greater than that obtained at low-temperature IR drying. Thus, for the polymer coating used to encapsulate sorghum seeds, only mild drying rate conditions should be applied to minimize cracking under moisture removal, thus preserving structural integrity of the gel-seed system.

ACKNOWLEDGEMENTS

The authors are grateful to CNPq (National Council for Research) and FAPITEC (Foundation for Research Support of Sergipe State) for the financial support received.

REFERENCES

1. G. Koçar and N. Civaş, "An Overview of Biofuels from Energy Crops: Current Status and Future Prospects," Renewable and Sustainable Energy Reviews, Vol. 28, 2013, pp. 900-916.
2. G. G. Cordeiro, G. M. de Resende, J. R. Pereira and N. D. Costa, "Effect of Saline Water and Soil Conditioner on Sugar Beet Yield in the Brazilian Semi-Arid Region," Horticultura Brasileira, Vol. 17, No. 1, 1999, pp. 39-41.
3. M. D. Kaya, G. Okc¸ M. Atak, Y. Çıkılı and O. Kolsarıcı, "Seed Treatments to Overcome Salt and Drought Stress during Germination in Sunflower (Helianthus annuus L.)," European Journal of Agronomy, Vol. 24, No. 4, 2006, pp. 291-296.
4. A. G. Taylor, P. S. Allen and M. K. Misra, "Seed Enhancements," Seed Science Research, Vol. 8, No. 21, 1998, pp. 245-256.
5. M. M. Prado, M. M. Ferreira and D. J. M. Sartori, "Drying of Seeds and Gels," Drying Technology, Vol. 24, No. 3, 2006, pp. 281-292.
6. S. Sarrocco, R. Raeta and G. Vannacci, "Seeds Encapsulation in Calcium Alginate Pellets," Seed Science and Technology, Vol. 32, No. 3, 2004, pp. 649-661.
7. J. Kajtna, U. Sebenik, M. Krajnc and J. Golob, "IR Drying of Water-Based Acrylic PSA Adhesives," Drying Technology, Vol. 26, No. 3, 2008, pp. 323-333.

8. G. S. V. Raghavan, T. J. Rennie, P. S. Sunjka, V. Orsat, W. Phaphuangwittayakul and P. Terdtoon, "Overview of New Techniques for Drying Biological Materials with Emphasis on Energy Aspects," Brazilian Journal of Chemical Engineering, Vol. 22, No. 2, 2005, pp. 195-201.

9. H. Kocabiyik and D. Tezer, "Drying of Carrot Slices Using Infrared Radiation," International Journal of Food Science and Technology, Vol. 44, No. 5, 2009, pp. 953- 959.

10. J. Seyed-Yagoobi and J. W. Wirtz, "An Experimental Study of Gas-Fired Infrared Drying of Paper," Drying Technology, Vol. 19, No. 6, 2001, pp. 1099-1112.

11. S. Le Person, J. R. Puiggali, M. Baron and M. Roques, "Near Infrared Drying of Pharmaceutical Thin Films: Experimental Analysis of Internal Mass Transport," Chemical Engineering and Processing, Vol. 37, No. 3, 1998, pp. 257-263.

12. D. Blanc, P. Laurent, J. F. Gerard and J. Andrieu, "Experimental Infrared Drying Study of a Model Water-Based Epoxy-Amine Painting Coated on Iron Support," Drying Technology, Vol. 15, No. 6-8, 1997, pp. 1787-179 9.

13. S. B. Pawar, P. S. R. Kumar, A. S. Mujumdar and B. N. Thorat, "Infrared-Convective Drying of Organic Pigments," Drying Technology, Vol. 26, No. 3, 2008, pp. 315-322.

14. P. Glouannec, P. Salagnac, H. Guézenoc and N. Allanic, "Experimental Study of Infrared-Convective Drying of Hydrous Ferrous Sulphate," Powder Technology, Vol. 187, No. 3, 2008, pp. 280-288.

15. A. R. Celma, S. Rojas and F. Lopez-Rodríguez, "Mathematical Modelling of Thin-Layer Infrared Drying of Wet Olive Husk," Chemical Engineering and Processing, Vol. 47, No. 9-10, 2008, pp. 1810-1818.

16. N. Boudhrioua, N. Bahloul, I. B. Ben and N. K. Slimen, "Comparison on the Total Phenol Contents and the Color of Fresh and Infrared Dried Olive Leaves," Industrial Crops and Products, Vol. 2, No. 2-3, 2009, pp. 412-419.

17. I. Das, S. K. Das and S. Bal, "Drying Kinetics of High Moisture Paddy Undergoing Vibration-Assisted Infrared (IR) Drying," Journal of Food Engineering, Vol. 95, No. 1, 2009, pp. 166-171.

18. C. Niamnuy, M. Nachaisin, N. Poomsa-ad and S. Devahastin, "Kinetics Modeling of Drying and Conversion/ Degradation of Isoflavones during Infrared Drying of Soybean," Food Chemistry, Vol. 133, No. 3, 2012, pp. 946- 952.

19. S. Eichler, O. Ramon and I. Ladyzhinski, "Collapse Processes in Shrinkage of Hydrophilic Gels during Dehydration," Food Research International, Vol. 30, No. 9, 1997, pp. 719-726.

20. G. Hashemi, D. Mowla and M. Kazemeini, "Moisture Diffusivity and Shrinkage of Broad Beans during Bulk Drying in an Inert Medium Fluidized Bed Dryer Assisted by Dielectric Heating," Journal of Food Engineering, Vol. 92, No. 3, 2009, pp. 331-338.

21. I. Bialobrzewski, M. Zielínska, A. S. Mujumdar and M. Makowski, "Heat and Mass Transfer during Drying of a Bed of Shrinking Particles—Simulation for Carrot Cubes Dried in a Spout-Fluidized-Bed Drier," International Journal of Heat and Mass Transfer, Vol. 51, No. 19, 2008, pp. 4704-4716.

22. Committee on Food Chemicals Codex, "Food Chemical Codex," 5th Edition, National Academies Press, Washington DC, 2004.

23. O. Fasina, B. Tyler, M. Pickard, G. Zheng and N. Wang, "Effect of Infrared Heating on the Properties of Legume Seeds," International Journal of Food Science and Technology, Vol. 36, No. 1, 2001, pp. 79-90.

24. J. Crank, "The Mathematics of Diffusion," 2nd Edition, Clarendon Press, Oxford, London, 1975.

25. B. A. Souraki and D. Mowla, "Axial and Radial Moisture Diffusivity in Cylindrical Fresh Green Beans in a Fluidized Bed Dryer with Energy Carrier: Modeling with and without Shrinkage," Journal of Food Engineering, Vol. 88, No. 1, 2008, pp. 9-19.

26. A. Arévalo-Pinedo and F. E. X. Murr, "Kinetics of Vacuum Drying of Pumpkin (Cucurbita maxima): Modeling with Shrinkage," Journal of Food Engineering, Vol. 76, No. 4, 2006, pp. 562-567.

27. R. Fletcher, "Pratical Methods of Optimization," 2nd Edition, Wiley, London, 2001.

28. C. Ratti, "Shrinkage during Drying of Foodstuffs," Journal of Food Engineering, Vol. 23, No. 1, 1994, pp. 91- 105.

29. L. Mayor and A. M. Sereno, "Modelling Shrinkage during Convective Drying of Food Materials: A Review," Journal of Food Engineering, Vol. 61, No. 3, 2004, pp. 373-386.

30. S. Pabis and M. Jaros, "The First Period of Convection of Vegetables and the Effect of Shape-Dependent Shrinkage," Biosystems Engineering, Vol. 81, No. 2, 2002, pp. 201- 211.

31. P. Perré and B. May, "The Importance of Considering Exchange Surface Area Reduction to Exhibit a Constant Drying Flux Period in Foodstuffs," Journal of Food Engineering, Vol. 54, No. 4, 2002, pp. 271-282.

32. G. P. Sharma, R. C. Verma and P. Pathare, "Mathematical Modeling of Infrared Radiation Thin Layer Drying of Onion Slices," Journal of Food Engineering, Vol. 71, No. 3, 2005, pp. 282-286.

33. A. K. Datta and H. Ni, "Infrared and Hot-Air-Assisted Microwave Heating of Foods for Control of Surface Moisture," Journal of Food Engineering, Vol. 51, No. 4, 2002, pp. 355-364.

34. I. Koç, R. I. Eren and F. K. Ertekin, "Modelling Bulk Density, Porosity and Shrinkage of Quince During Drying: The Effect of Drying Method," Journal of Food Engineering, Vol. 85, No. 3, 2008, pp. 340-349.

CHAPTER 2

Mass Transfer in Amperometric Biosensors Based on Nanocomposite Thin Films of Redox Polymers and Oxidoreductases

Michael V. Pishko [1], Alexander Revzin [2] and Aleksandr L. Simonian [2]

[1]Department of Chemical Engineering, The Pennsylvania State University, 104 Fenske Laboratory, University Park, PA 16802-4400

[2]Texas A&M University, Department of Chemical Engineering, College Station, TX 77843-3122

ABSTRACT

Mass transfer in nanocomposite hydrogel thin films consisting of alternating layers of an organometallic redox polymer (RP) and oxidoreductase enzymes was investigated. Multilayer nanostructures were fabricated on gold surfaces by the deposition of an anionic self-assembled monolayer of 11-mercaptoundecanoic acid, followed by the electrostatic binding of a cationic redox polymer, poly[vinylpyridine Os(*bis*-bipyridine)2Cl- co-allylamine], and an anionic oxidoreductase. Surface plasmon resonance spectroscopy, Fourier transform infrared external reflection spectroscopy (FTIR-ERS), ellipsometry and electrochemistry were employed to characterize the assembly of these nanocomposite films. Simultaneous SPR/electrochemistry enabled real time observation of the assembly of sensing components, changes in film structure with electrode potential, and the immediate, *in situ*electrochemical verification of substrate-dependent current upon the addition of enzyme to the multilayer structure. SPR and FTIR-ERS studies also showed no desorption of polymer or enzyme from the nanocomposite structure when stored in aqueous environment occurred over the period of three weeks, suggesting that decreasing in substrate sensitivity were due to loss of enzymatic activity rather than loss of film compounds from the nanostructure.

KEYWORDS

Biosensor; Redox; Nanocomposite; Hydrogel; Enzyme

1. INTRODUCTION

The electrostatic layer-by-layer assembly of nanocomposite films composed of alternating layers of oppositely charged molecules has been a subject of numerous publications.[1,2,3,4,5,6, 7,8,9,10, 11,12,13,14,15,16] A promising application of this assembly approach is in the area of biosensor fabrication. Layer-by-layer assembly represents a potentially reproducible method to control the architecture of the nanometer-scale biosensing films. In addition, electrostatic assembly of sensing layers is well-suited for mediated electrochemical sensing approaches using charged macromolecular electron acceptor/donors and oxidoreductases. First proposed and implemented by Heller and co-workers,[17] this method entails the electrostatic binding between cationic redox polymers and anionic oxidoreductase enzymes which results in the transfer of electrons from the enzyme via the redox polymer to the electrode surface.[4,6,8,13,14] Thus, layer-by-layer assembly of oppositely charged redox polymers and oxidoreductase enzymes takes advantage of this method. Biosensors utilizing various electrostatically complexed redox polymer and enzyme couples have been successfully used to detect glucose, lactate and other analytes. While much attention has been directed to the application of the electrostatically assembled biosensors, few studies have sought to characterize the layer-by-layer electrostatic assembly process or mass transfer within these films. Calvo and co-workers used QCM, AFM, and electrochemical analysis to investigate the assembly process,[4,6] while others employed XPS and SPR for the same purpose.[6,13]

In this manuscript we described mass transfer and hydrogel swelling in electrostatically assembled nanocomposite thin films containing redox polymers and oxidoreductases. Surface plasmon resonance spectroscopy (SPR), Fourier transform infrared – external reflection spectroscopy (FTIR-ERS), and ellipsometry were used to characterize film structure. Cyclic voltammetry was used to measure analyte dependent electrode currents that are a function of analyte mass flux in the thin films. While emphasis in this study was placed on the characterization of nanocomposite thin films containing GOX, we have also studied nanocomposite structures containing POD and LOX.

2. MATERIALS AND METHODS

2.1 Reagents

Glucose oxidase (GOX, EC 1.1.3.4, Type X-S, 128 units/mg solid from *Aspergillus niger*) and lactate oxidase (LAX, 35 units/mg solid, from *Pediococcus* species) were obtained from the Sigma- Aldrich Chemical Co. (St. Louis, MO). Pyruvate oxidase (PYX, EC 1.2.3.3, MW

260 kDa) was purchased from Toyobo America, Inc. (New York, NY). 11-mercaptoundecanoic acid (MUA) was obtained from the Aldrich Chemical Co. (Milwaukee, WI). All reagents, unless otherwise stated, were used as received. Phosphate buffered saline (PBS: 10 mM phosphate, 138 mM NaCl, and 2.7 mM KCl) was in 18 MΩ℘cm distilled, deionized water (E-pure, Barnstead). Silver epoxy was obtained from Epoxy Technologies (Billerica, MA). Silicon wafers coated with Cr (20 Å) and Au (500 Å) were purchased from Lance Goddard Associates (Foster City, CA).

Polycationic redox polymer, poly[vinylpyridine Os(*bis*-bipyridine)2Cl-co-allylamine] (PVP-Os-AA) was synthesized according to the procedures described previously.[8] The structure of this polymer is shown inFigure 1. In its reduced and oxidized forms, the oxidation state on the organometallic osmium complex is +2/+3. However, the overall charge on the complex is +1/+2 in its reduced and oxidized states, respectively.

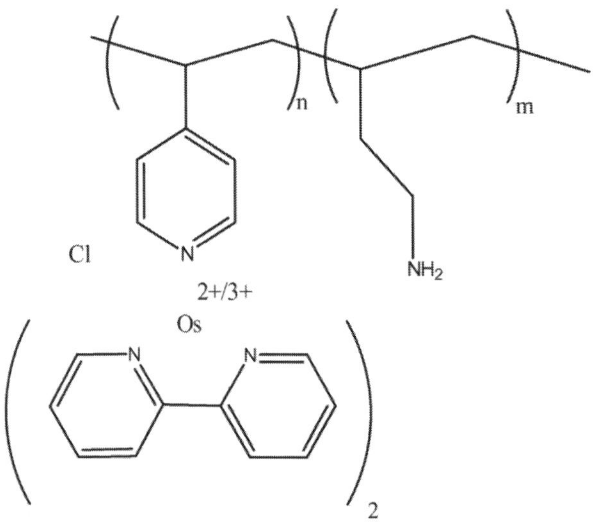

Figure 1. Structure of the polycationic redox polymer used in these studies.

2.2 Equipment

The Spreeta sensor, a miniature (approximately 7 g), fully integrated surface plasmon resonance device produced by Texas Instruments, was described in detail elsewhere.[18,19] Briefly, the sensor possesses a light-emitting diode with a wavelength of 825 nm, a polarizer, a thermistor, and two 128- pixel silicon photodiode arrays (EOC 1401). All components are seated on a small printed circuit board and encapsulated with clear, optical epoxy (index of refraction of 1.52), molded to have optically flat surfaces. Because the system was fully encapsulated, no optical alignment is necessary. The molded epoxy is shaped in the form of the Kretschmann geometry prism to facilitate the excitation and detection of surface plasmon resonance. The surface plasmon resonance glass sensing side of the sensor was modified with 10 to 20 Å chromium adhesion layer evaporated onto the surface, followed by the evaporation of 500 Å of gold onto the chromium.

A small flow cell was attached to the sensor to enable delivery of solutions across the sensing surface. In the electrochemical experiments, the gold surface of the SPR substrate also serves as the working electrode. Electrical contact with the gold surface was realized by attaching a wire via conductive silver epoxy; the contacted area was excluded from the flow cell. An additional Ag/AgCl reference and Pt counter electrodes were introduced at the top of the flow cell to create a three- electrode configuration.

Each experiment was performed with a new Au surface. Before starting an experiment, the gold surface of the SPR substrate was thoroughly cleaned using a cotton tip, dipped in 0.1% Triton and followed by immersion in 0.12 N NaOH to remove organic residues on the Au surface. The flow cell was assembled and system was flushed first with ethanol, then with DI water, 0.1% Triton and 0.12 N NaOH solution followed by DI water again. For all experiments the flow cell was equilibrated in running PBS buffer until a stable baseline SPR signal was obtained. All data was acquired using refractive index (RI) temperature compensation and each data point was the average of 3 - 5 measurements. Experiments were carried out at room temperature (20-21° C).

Equipment for the electrochemical analysis included a CV-50W Voltammetric Analyzer (Bioanalytical Systems, West Lafayette, IN). The electrochemical apparatus was controlled and data acquired using a Dell Pentium PC. Cyclic voltammetry from 0 to 500 mV (vs. Ag/AgCl electrode), at the scan rate of 20 mV/s, was used to measure the substrate-dependent current produced by the redox polymer/ oxidoreductase thin films challenged with varying concentrations of enzyme substrate. Electrodes were tested in a cell of volume approximately 1 mL containing 0.1 M PBS degassed with N2 (15 min. prior to start of experiment, for duration of experiment) and also containing Pt counter and Ag/AgCl reference electrodes. Degassing with N2 agitated the solution created oxygen-free environment.

Ellipsometry was performed using a Gaertner LSE Stokes Ellipsometer (Gaertner Scientific Corp., Chicago, IL). Film thickness and refractive indices were calculated using Gaertner software. All measurements were made using a He/Ne (633.8 nm) laser and a 70.00 +/- 0.02o angle of incidence.

3. RESULTS AND DISCUSSIONS

We have formed redox polymer and enzyme complexes through an electrostatic layer by layer deposition scheme. This scheme for sensor development minimizes the use of such components as retaining membranes or thick polymer matrices, resulting in an increased surface density of the primary sensing components, namely the enzyme and redox polymer, and thus may potentially improve sensor sensitivity and reduce sensor response time. Minimizing the thickness of the sensing films was achieved by keeping the thickness of the sensing layers well under 100 Å through repeated attachment of single redox polymer/enzyme monolayers.

3.1 Thin film assembly and structure

The scheme used for fabrication of these sensors is based upon electrostatic attraction between a polycationic osmium redox polymer and a polyanionic enzyme, GOX, LAX or PYX. First, the electrode surface was functionalized with MUA where the thiol end group was chemisorbed to gold. The carboxylic acid end group of MUA presented the solution-electrode interface with a net negative charge, since with a pKa of 6.5 the acid group is at least partially deprotonated at a pH of 7.4.[20] Polymers modified with organometallic osmium complexes are known to be polycationic[8,14] at neutral pH and can thus be electrostatically bound to the negatively functionalized surface. Subsequent electrostatic binding of anionic GOX with an isoelectric point of ~ 3.0 or LAX with isoelectric point ~ 4.6[8,14] was then possible to the positively charged redox polymer layer. This process was then repeated as desired to deposit multiple layers. A schematic of a final nanocomposite structure is shown in Figure 2.

Figure 3 shows changes in refractive index at the SPR Au surface with time for two binding steps, the initial deposition of RP on the MUA monolayer and the initial deposition of GOX on the RP layer.

The rapid signal increase in the first few seconds after introduction of RP may be attributed to the change in the bulk refractive index of solution.[19,20] This is followed by a gradual increase that represents the adsorption of material from solution onto the anionic MUA surface. When the surface coverage of RP has reached equilibrium, a plateau in the SPR signal is observed (see Figure 3). At this point, the maximum concentration of polyelectrolyte has adsorbed onto the surface. Once a wash solution is introduced again into the flow-through cell, another sharp change (drop) in SPR response indicates change in the bulk refractive index of solution. This is followed by the desorption of the loosely bound ligand from the charged surface. Thus, a difference between initial and final RI represents the change in RI due to surface bound RP. As can be seen from the SPR signal, this amount of adsorbed RP does not wash from the surface as each RP chain is held in place by multivalent contacts between the polymer chain and the MUA monolayer. Figure 3also shows a deposition cycle for an oxidoreductase, GOX. Because GOX is a polyanion, it readily adsorbs to the RP layer and again is held in place by multivalent contacts. Continuous SPR monitoring over a 3 week period indicated that these nanocomposite films did not dissociate upon continuous exposure to water. Similar results both in terms of nanocomposite assembly and structural stability were obtained for LAX and PYX, both of which are polyanions at neutral pH. FTIR-ERS confirmed the binding of all enzymes as indicated by the appearance of the amide I and amide II absorption peaks associated with the polypeptide backbone. For all the enzymes, one would expected that the enzyme is highly intercalated with the redox polymer layer, given the highly charged nature of each macromolecule in the multilayer structure.[7,13,15,16,21,22]

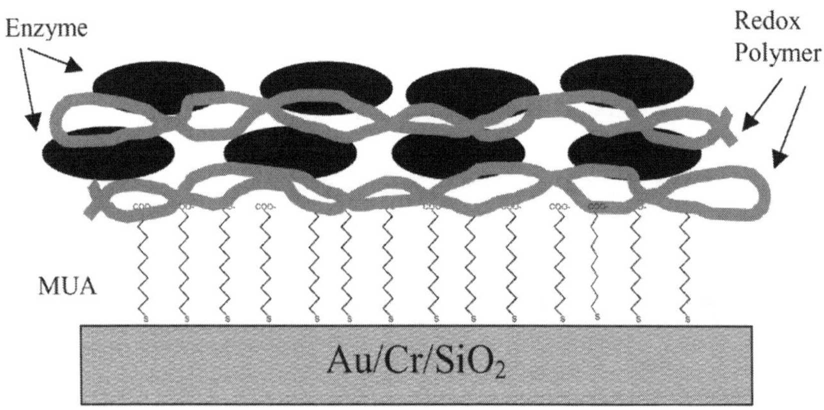

Figure 2. Schematic of the nanocomposite structure created by the layer-by-layer deposition of alternating polyanionic and polycationic films on gold.

The SPR signal can be directly related to the surface concentration of proteins with a shift of θ_{SPR} by 1° degree corresponding to the deposition of 10 ng/mm^2 of protein.[23] Because the SPR signal can be alternatively recorded in RI units or angle θ_{SPR}, the surface concentration of enzyme for each deposition step was easily found using the conversion factor of 7x10^{-3} RIU/degree. A surface coverage of 2.4 ng/mm^2 in the first GOX assembly step and 1.9 ng/mm^2 for the second GOX assembly step were determined. These compare favorably with the 3.1 ng/mm^2 obtained by Hodak and co- workers in a quartz crystal microbalance study of GOX deposition onto a positively charged gold surface.[6] Surface coverages for LOX and PYX were estimated to be 1.29 ng/mm^2 and 10.7 ng/mm^2 respectively.

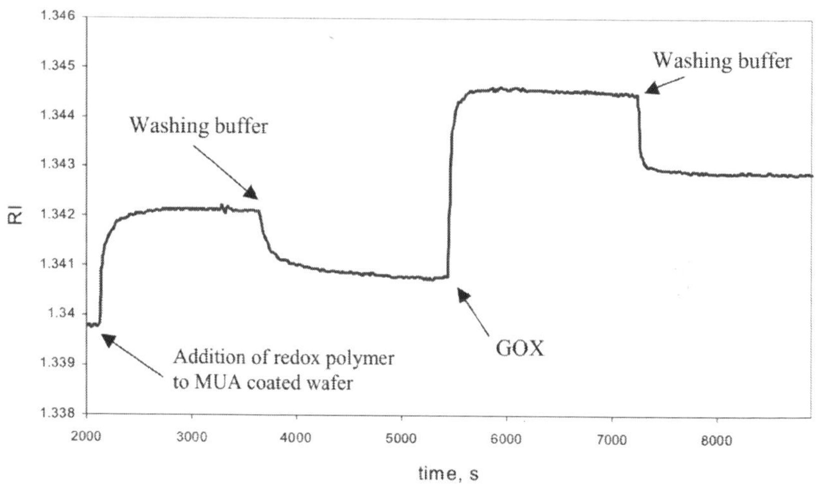

Figure 3. Changes in refractive index with time as measured by SPR during the assembly of a nanocomposite thin film consisting of MUA and one RP/GOX bilayers. The concentrations of RP and GOX were 10 mg/mL in PBS, flow rate 80 µL/min.

Assuming the density of protein to be 1.3 g/cm3 and a densely packed layer,[12] the thickness of the first deposited enzyme layer was approximately 19 Å. While these results do not represent true height of the adsorbed enzyme because surface coverage is not 100 %, it may still indicate presence of a monolayer of protein on the positively charged surface. However, given known crystallographic dimensions of GOX,[24] it is highly unlikely that all the adsorbed protein is GOX. This result is not unexpected. All enzyme formulations are contaminated with other proteins and protein fragments of the enzyme of interest. For example, GOX Type XS from Sigma-Aldrich is known to be contaminated with catalase and invertase. In addition, the formulation may likely contain lytic and denatured fragments of GOX and other proteins. Thus the adsorbed protein layer may be a mixture of proteins possessing net negative charge.

The maximum surface coverage (in ng/mm^2) for a densely packed enzyme monolayer may be found using the following formula[25]:

$$P_{ideal} \sim \frac{10^{21} MW}{\pi ab N_a} \quad (1)$$

where MW is molecular weight (18.6 kDa for GOX, 260 kDa for POD and 80 kDa for LOX), Na is Avogadro's number and πab is an ellipsoid surface projection of the globular enzyme, with *a* and *b* being the radii of the short and long axis of the ellipse. In case of PYX and LOX, crystallographic data does not exist and the ellipsoid surface projection was simplified to πr^2 where r is a hydrodynamic radius. Taking GOX dimensions of 30 Å and 40 Å for *a* and *b* respectively,[24] P_{ideal} was calculated to be 8.19 ng/mm^2. In case of PYX, taking into account 50 Å hydrodynamic radius, P_{ideal} was found to be 5.5 ng/mm^2. Surface coverage was then found by means of the following relationship:

$$\text{Surface Coverage} = \left(\frac{P_{expected}}{P_{ideal}}\right) \times 100\% \quad (2)$$

Surface coverage estimates for the first two GOX layers were 30% and 23% respectively and are reasonable considering that the maximum obtainable coverage for the irreversible adsorption of protein is hypothesized to 55%.[26,27] Such low surface coverage is thought to occur because of the steric hindrance and electrostatic repulsion between adsorbing protein molecules.[26] Another explanation may be the fact that lyophilized enzyme powders are not pure (as mentioned earlier) and thus the measurement techniques report an average measurement for all the proteins adsorbed. Thus, one could expect competitive protein adsorption onto the redox polymer surface, which may also account for a lower surface coverage for GOX.

While lack of crystallographic data for LOX does not permit one to approximate surface coverage, one could infer the formation of a sub-monolayer of enzyme from a relatively weak surface coverage. On the other hand, PYX surface coverage is estimated to be 194 %. This result may indicate surface aggregation of the less stable PYX tetramer[28,29] and the formation of multiple enzyme layers in a single PYX deposition step.

As measured by ellipsometry for a dry film, the thickness was ~10 Å for a single GOX layer deposited on a RP layer. This strongly suggests that these nanocomposite thin films are highly swollen by water. One would expect this to be true, given that the nanocomposite films are formed from water soluble polyelectrolytes. Water penetration is also a necessary feature for aqueous electrochemistry in redox polymer thin films. As RP is oxidized, the oxidation state of Os changes from +2 to +3 and thus the charge on the complex changes from +1 to +2. Electroneutrality must be maintained and thus counter ions must move into the film along with associated water molecules of hydration.

This change in film hydration upon the oxidation and reduction of RP is clearly demonstrated in Figure 4. Here, a nanocomposite film comprised on two RP/GOX bilayers was oxidized and reduced using cyclic voltammetry while simultaneously measuring refractive index using SPR spectroscopy. Upon oxidation at 500 mV(Ag/AgCl), Os sites on RP change from +1 to +2, and counter-ions will move into the film to maintain Electroneutrality. Water associated with these counter-ions will also diffusing in. This flux of water and ions into the nanocomposite film is evident from the decrease in RI as shown in Figure 3 (i.e. the RI in the film moves closer to that of water). When RP is reduced at 0 mV, the charge in the polymer decreases to +1, water and counter-ions move out of the film, and the film collapses, resulting in an increase in RI (i.e. the RI in the film moves closer to that for a condensed polymer). Because nanocomposite films created from layers of polyelectrolytes have been shown to be nanoporous, the expansion and collapse of the thin film with potential may also represent a change in pore structure and volume at the nanometer level.[16] Fery and colleagues observed changes in nanopore structure based on changes in salt concentration of layer- by-layer polyelectrolyte nanocomposites. One may hypothesize that changes in salt concentration as produced by changes in potential may produce a similar effect.

3.2 Electron transfer and mass transfer at the sensor surface

The oxidoreductase-catalyzed reaction scheme with Os sites on RP serving as electron relays between the enzyme and electrode surface may be expressed as follows, using glucose oxidase as an example:

β-D-glucose + GOX(FAD) → Gluconolactone + GOX($FADH_2$)

GOX($FADH_2$) + 2 M^{+2} → GOX(FAD) + 2 M^{+1}

$M^{+2} + M^{+1} \rightarrow M^{+1} + M^{+2}$

2 $M^{+1} \rightarrow 2e^- + 2H^+ + 2M^{+2}$

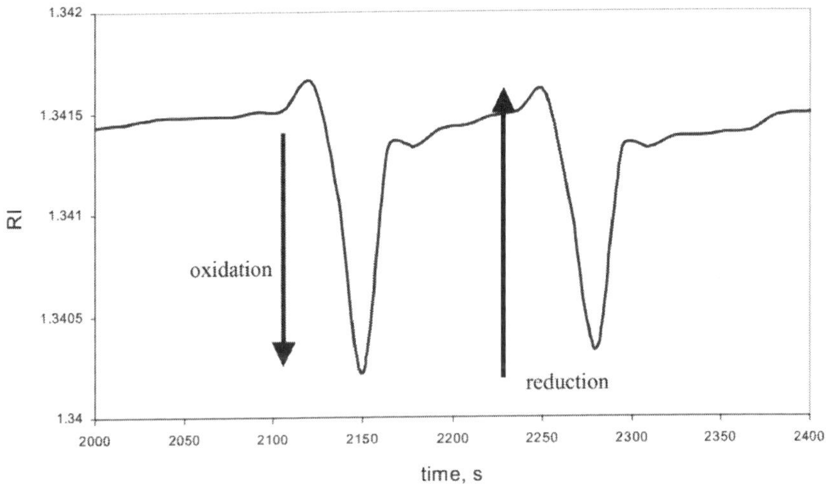

Figure 4. Changes in refractive index with time as measured by SPR during the oxidation and reduction of a nanocomposite thin film consisting of MUA and one RP/GOX bilayers.

where GOX(FAD) and GOX (FADH2) represent the oxidized and reduced forms, respectively, of the enzyme's cofactor – flavin adenine dinucleotide (FAD) and M represents an Os redox site of the polymer in its oxidized (+2) and reduced (+1) forms. Electrons are thus transferred from the enzyme to the redox polymer, shuttled between the redox sites in a self-exchange reaction until being transferred to an electrode surface that is poised positive of the formal potential of the polymer. As a result, a current that is proportional to substrate concentration is created. In the absence of mass transfer limitations, the concentration ($i = (i_{max}[s]/([s] + Kappm))$) dependence of the current is directly related to the Michaelis-Menton kinetics of catalysis for the enzyme, where [s] is the concentration of substrate (e.g. glucose), i_{max} is the maximum current possible at high substrate concentrations (analogous to the limiting enzyme turnover rate), and Kappm is the apparent Michaelis-Menton constant for the enzyme.

Figure 5 shows the concentration dependence of the current density for a nanocomposite thin film modified electrode comprised of two RP/GOX bilayers. The current density was roughly linear in the concentration range from 0 to 20 mM glucose, a result that was not anticipated given that K_m for GOX is 6 mM[30] and that the film was only several nanometers thickness, making mass transfer limitations unlikely. Nanocomposite films containing LAX and PYX show similar behavior. Figure 5 also shows the anticipated concentration dependence of the current in the absence of mass transfer limitations. As can be seen, mass transfer limitations do exist in these nanocomposite films, a hypothesis supported by several studies. Heller and colleagues constructed micrometer scale polyelectrolyte thin films that limited glucose mass transfer in implanted sensors.[31] Others have observed that polyelectrolyte nanocomposite thin films are highly compact, have a nanoporous structure and can limit the diffusion of small molecules such as ferricyanide.[15,16,22] Given that the nanocomposite films described here do undergo significant structural changes upon oxidation and reduction, these films are likely very compact and may posses nanoporosity as seen in other polyelectrolyte thin films.

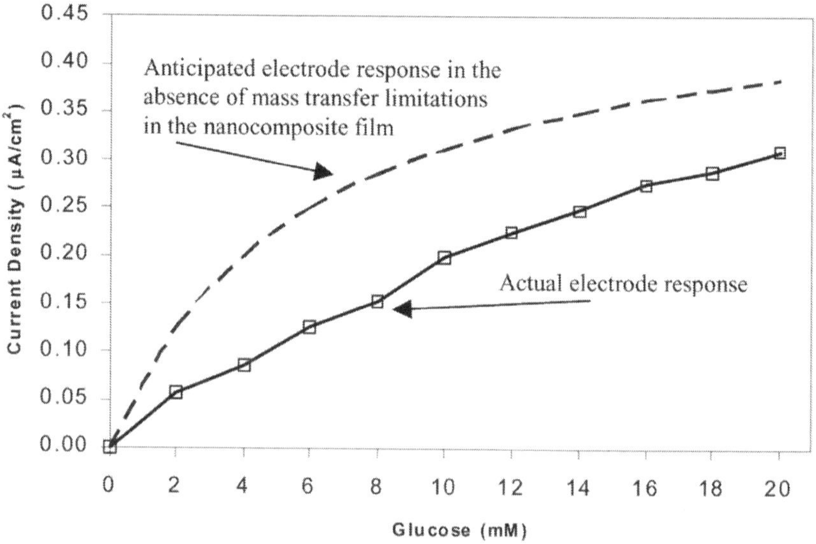

Figure 5. Concentration dependence of the electrode current density for a nanocomposite thin film comprised of two RP/GOX bilayers. The solid line shows the actual response of a nanocomposite thin film while the dash line shows the anticipated response in the absence of mass transfer limitations in the film. Currents were acquired using cyclic voltammetry (0 to 500 mV, 20 mV/s), with the current at 500 mV used in the current density calculation.

4. CONCLUSIONS

In conclusion, we have used SPR and ellipsometry to investigate the layer-by-layer assembly of nanocomposite films on a gold surface. Such investigation is an important step toward better understanding of assembly and properties of enzyme containing nanocomposite thin films. The surface coverages of the deposited GOX, LOX and POD enzymes were estimated by means of SPR spectroscopy. Combined electrochemistry and SPR was used to observe structural changes in the nanocomposite films resulting from the movement of water and ions into and out of the film. The concentration dependence of the current for a glucose oxidase-containing nanocomposite film confirmed that the film structure does impose mass transfer limitations, suggesting that it is highly compact and may be nanoporous.

ACKNOWLEDGEMENTS

Authors would like to thank Texas Instruments Incorporated and Dr. James Elkind for providing the TISPR-1 sensor and Experimenter's Kit. We would also like to thank Dr. Richard Crooks from the Department of Chemistry at Texas A&M University and his group members for the use of FTIR-ERS instrument. Financial support was provided by the National Science Foundation (CTS-9875372) and the United States Department of the Army and the National Medical Technology Testbed, Inc. (DAMD17-97-2-7016) for funding. MVP acknowledges the support of the Alfred P. Sloan Foundation through a research fellowship. "This work does not necessarily reflect the position or the policy of the government or NMTB, and that no official endorsement should be inferred."

REFERENCES

1. Anicet, N.; Bourdillon, C.; Moiroux, J.; Saveant, J.-M. Electron Transfer in Organized Assemblies of Biomolecules. Step-by-Step Avidin/Biotin Construction and Dynamic Characteristics of a Spatially Ordered Multilayer Enzyme Electrode. *J. Phys. Chem. B* **1998**, *102*, 9844–9849.

2. Anzai, J.; Kobayashi, Y.; Nakamura, N.; Nishimura, M.; Hoshi, T. Layer-by-layer construction of multilayer thin films composed of avidin and biotin-labeled poly(amine)s. *Langmuir* **1999**, *15*, 221–226.

3. Bacha, S.; Montagne, M.; Bergel, A. Modeling Mass Transfer with Enzymatic Reaction in Electrochemical Multilayer Microreactors. *AIChE J.* **1996**, *42*, 2967–2976.

4. Calvo, E.; Battaglini, F.; Danilowicz, C.; Wolosiuk, A.; Otero, M. Layer-by-layer electrostatic deposition of biomolecules on surfaces for molecular recognition, redox mediation and signal generation. *Faraday Discuss* **2000**, *116*, 47–65.

5. Clark, S. L.; Hammond, P. T. Engineering the Microfabrication of Layer-by-Layer Thin Films. *Adv. Mater.* **1998**, *10*, 1515–1519.

6. Hodak, J.; Etchenique, R.; Calvo, E. J.; Singhal, K.; Bartlett, P. N. Layer-by-Layer Self-Assembly of Glucose Oxidase with a Poly(allylamine)ferrocene Redox Mediator. *Langmuir* **1997**, *13*, 2708–2716.

7. Hou, S.; Yang, K.; Fang, H.; Chen, H. Amperometric glucose enzyme electrode by immobilizing glucose oxidase in multilayers on self-assembled monolayers surface. *Talanta* **1998**, *47*, 561–567.

8. Sirkar, K.; Revzin, A.; Pishko, M. V. Glucose and lactate biosensors based on redox polymer/oxidoreductase nanocomposite thin films. *Anal. Chem.* **2000**, *72*, 2930–2936.

9. Sun, J.; Wu, T.; Sun, Y.; Wang, Z.; Zhang, X.; Shen, J.; Cao, W. Fabrication of a covalently attached multilayer via photolysis of layer-by-layer self-assembled films containing diazo-resins. *Chem. Commun.* **1998**, 1853–1856.

10. Sun, Y.; Sun, J.; Zhang, X.; Sun, C.; Wang, Y.; Shen, J. Chemically modified electrode via layer- by-layer deposition of glucose oxidase (GOD) and polycation-bearing Os complex. *Thin Solid Films* **1998**,*327*, 730–733.

11. Anzai, J. K.; Suzuki, Y.; Takeshita, H.; Chen, Q.; Osa, T.; Hoshi, T.; Du, X. Enzyme sensors prepared by layer-by-layer deposition of

enzymes on a platinum electrode through avidin-biotin interaction.*Sensors and Actuators* **1998**, *52*, 3–9.

12. Lvov, Y.; Ariga, K.; Ichinose, I.; Kunitake, T. Assembly of multicomponent protein films by means of electrostatic layer-by-layer adsorption. *J. Am. Chem. Soc.* **1995**, *117*, 6117–6123.

13. Laurent, D.; Schlenoff, J. Multilayer Assemblies of Redox Polyelectrolytes. *Langmuir* **1997**, *13*, 1552–1557.

14. Revzin, A.; Sirkar, K.; Simonian, A.; Pishko, M. Glucose, Lactate, and Pyruvate Biosensor Arrays Based on Redox Polymer/Oxidoreductase Nanocomposite Thin Films Deposited on Photolithographically Patterned Gold Microelectrodes. *Sensors and Actuators* **2002**, *81*, 359–368.

15. Dubas, S.; Schlenoff, J. Swelling and smoothing of polyelectrolyte multilayers by salt. *Langmuir* **2001**,*17*, 7725–7727.

16. Fery, A.; Scholer, B.; Cassagneau, T.; Caruso, F. Nanoporous thin films formed by salt-induced structural changes in multilayers of poly(acrylic acid) and poly(allylamine). *Langmuir* **2001**, *17*, 3779–3783.

17. Pishko, M. V.; Katakis, I.; Lindquist, S.-E.; Ye, L.; Gregg, B. A.; Heller, A. Direct Electrical Communication between Graphite Electrodes and Surface Adsorbed Glucose Oxidase/Redox Polymer Complexes. *Angew. Chem. Intl. Ed.* **1990**, *29*, 82.

18. Melendez, J.; Carr, R.; Bartholomew, D.; Kukanskis, K.; Elkind, J.; Yee, S.; Furlong, C.; Woodbury, R. Spreeta SPR instrument. *Sensors and Actuators B* **1996**, *35-36*, 212–216.

19. Elkind, J.; Stimpson, D.; Strong, A.; Bartholomew, D.; Melendez, J. Spreeta SPR instrument. *Sensors and Actuators B* **1999**, *54*, 182–190.

20. Franchina, J.; Lackowski, W.; Dermody, D.; Crooks, R.; Bergbreiter, D.; Sirkar, K.; Pishko, M. Entrapment of a biorecognition molecule

in a weak-acid, polyelectrolyte hyperbranched thin film on gold. *Anal. Chem.* **1999**, *71*, 3133–3139.

21. Hammond, P. T.; Whitesides, G. M. Formation of Polymer Microstructures by Selective Deposition of Polyion Multilayers Using Patterned Self-Assembled Monolayers as a Template. *Macromolecules* **1995**,*28*, 7569–7571.

22. Sukhorukov, G. B.; Schmitt, J.; Decher, D. Reversible Swelling of Polyanion/Polycation Multilayer Films in Solutions of Different Ionic Strength. *Ber. Bunsenges. Phys. Chem.* **1996**, *6*, 948–953.

23. Stenberg, E.; Persson, B.; Roos, H.; Urbaniczky, C. *J. Colloid Interf, Sci.* **1991**, *143*, 513–526.

24. Hecht, H.; Kalisz, H.; Hendle, J.; Schmid, R.; Schomburg, D. Crystal structure of glucose oxidase from Aspergillus niger refined at 2.3 A resolution. *J. Mol. Biol.* **1993**, *229*, 153–172.

25. Lahiri, J.; Isaacs, L.; Tien, J.; Whitesides, G. A Strategy for the Generation of Surfaces Presenting Ligands for Studies of Binding Based on an Active Ester as a Common Reactive Intermediate: A Surface Plasmon Resonance Study. *Anal. Chem.* **1999**, *71*, 777–790.

26. Yuan, Y.; Oberholzer, M.; Lenhoff, A. *Colloids and Surfaces* **2000**, *165*, 125–141.

27. Jin, X.; Wang, N.; Tarjus, G.; Talbot, J. *J. Phys. Chem.* **1993**, *97*, 4256–4258.

28. Gajovic, N.; Habermuller, K.; Warsinke, A.; Schuhmann, W.; Scheller, F. A pyruvate oxidase electrode based on electrochemically deposited redox polymer. *Electroanalysis* **1999**, *11*, 1377–1383.

29. Gajovic, N.; Binyamin, G.; Warsinke, A.; Scheller, F.; Heller, A. Operation of a miniature redox hydrogel-based pyruvate sensor in

undiluted deoxygenated calf serum. *Anal. Chem.* **2000**, *72*, 2963–2968.

30. Frederick, K. R.; Tung, J.; Emerick, R. S.; Masiarz, F. R.; Chamberlain, S. H.; Vasavada, A.; Rosenberg, S.; Chakraborty, S.; Schopter, L. M.; Massey, V. Glucose Oxidase from Aspergillus Niger. Cloning, Gene Sequence, Secretion from Sacharomyces Cerevisiae and Kinetic Analysis of a Yeast-Derived Enzyme. *J. Biol. Chem.* **1990**, *265*, 3793–3802.

31. Chen, T.; Friedman, K.; Lei, I.; Heller, A. In situ assembled mass-transport controlling micromembranes and their application in implanted amperometric glucose sensors. *Anal. Chem.* **2000**,*72*, 3757–3763.

CHAPTER 3

Liquid- and Gas-Phase Diffusion of Ferrocene in Thin Films of Metal-Organic Frameworks

Wencai Zhou, Christof Wöll and Lars Heinke

Institut für Funktionelle Grenzflächen (IFG), Karlsruher Institut für Technologie (KIT), Hermann-von-Helmholtz-Platz 1, Eggenstein-Leopoldshafen 76344, Germany

ABSTRACT

The mass transfer of the guest molecules in nanoporous host materials, in particular in metal-organic frameworks (MOFs), is among the crucial features of their applications. By using thin surface-mounted MOF films in combination with a quartz crystal microbalance (QCM), the diffusion of ferrocene vapor and of ethanolic and hexanic ferrocene solution in HKUST-1 was investigated. For the first time, liquid- and gas-phase diffusion in MOFs was compared directly in the identical sample. The diffusion coefficients are in the same order of magnitude (~10^{-16} $m^2 \cdot s^{-1}$), whereas the diffusion coefficient of ferrocene in the empty framework is roughly 3-times smaller than in the MOF which is filled with ethanol or *n*-hexane.

KEYWORDS

metal-organic frameworks; diffusion; mass transfer; gas phase; liquid phase

1. INTRODUCTION

Metal-organic frameworks (MOFs) are crystalline, nanoporous hybrid materials, which attract a considerable amount of attention due to their unique properties, such as their very large specific surface area and their enormous variety [1,2]. In addition to this, MOF properties can be varied over wide ranges in a rational manner, resulting in tunable structures. This enables many interesting applications, such as the storage and separation of gases, as sensors as well as in catalysis [3,4,5]. For all these applications, the interaction of the guest molecules in the pores with the framework is crucial. Therefore, the transport properties of the guest molecules are among the key properties of MOFs. The diffusion of various gas molecules in different MOF structures has been intensively studied, e.g., references [6,7,8,9,10]. On the other hand, the diffusion of the guest molecules from the liquid phase has only been investigated in a few publications [11,12,13]. So far, a comparison of the mass transfer properties in the MOF pores in the liquid and the gas phase is lacking.

Here, we used thin films of MOFs in combination with a quartz crystal microbalance (QCM) to investigate the uptake from the liquid and from the gas phase. The MOF films are prepared on a QCM sensor in a layer-by-layer fashion employing liquid-phase epitaxy. These thin MOF films are referred to as surface-mounted MOFs, SURMOFs

[14,15,16]. Various surfaces can be modified to direct and improve the SURMOF growth on top, e.g., several metal surfaces can be functionalized with a self-assembled monolayer (SAM) [17] of organic molecules with a thiol anchor group and a functional head group. The SURMOF crystal orientation is determined by the functionalization of the substrate surface [18], and their thickness can be directly controlled by the number of synthesis cycles. The SURMOF morphology is usually smooth and homogenous [19] and even transparent thin films can be prepared [20]. In addition to this, the SURMOF growth process may hinder the interpenetration of the MOF [21] and it enables hetero-epitaxy, *i.e.,* MOF-on-MOF [22,23]. SURMOF films in combination with QCM were used to investigate the diffusion of vapor molecules such as pyridine [24], cyclohexane [25] or ferrocene [26] in MOFs. On the other hand, the QCM was also used to investigate the SURMOF growth in the (alternating) synthesis solutions, *i.e.,* in the liquid phase [27]. Here, we combine both approaches and investigate the uptake of a probe molecule, ferrocene, from the gas phase,*i.e.,* from ferrocene vapor in a carrier gas, and from the liquid phase, *i.e.,* from ethanolic or hexanic ferrocene solutions, by the identical SURMOF sample of type HKUST-1, see Figure 1.

2. EXPERIMENTAL SECTION

For recording the mass changes and the uptake of the guest molecules by the SURMOF, a QCM from Q-Sense is used, which works at a resonance frequency of about 5 MHz. The investigated SURMOFs of type HKUST-1 [28] were prepared on gold-coated QCM sensors, whose surfaces were modified with an 11-mercaptoundecanol (MUD) SAM, in a layer-by-layer fashion by using the spray method [29]. By using the spray method, which enables upscaling to coating large areas, the ethanolic solutions of both MOF-components, *i.e.,* of the metal complexes (copper (ii) acetate) and of the organic linker molecules (1,3,5-benzenetricarboxylic acid), are alternatively sprayed on the substrate surface. The X-ray diffractogram of the prepared HKUST-1 SURMOF shows crystalline oriented growth, see Figure 2.

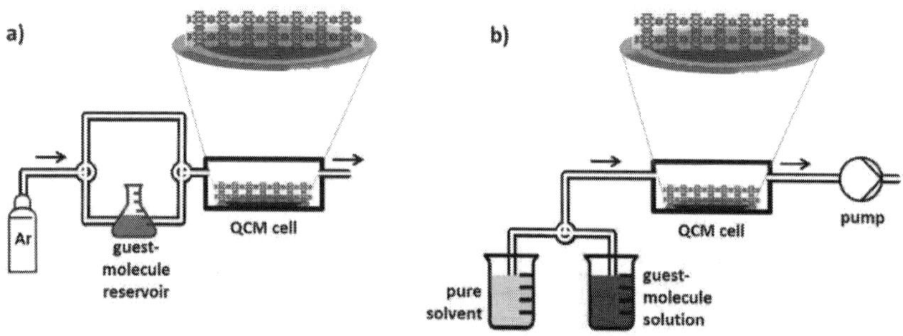

Figure 1. Setup of uptake experiments from gas and liquid phase. (**a**) A gas flow system where the carrier gas (here: argon) flows through the quartz crystal microbalance (QCM) cell, *i.e.*, over the QCM sensor, is used for uptake experiments from the gas phase. The carrier gas flow can be enriched with the vapor of the guest molecules by passing through (or over) the reservoir of the guest molecules; (**b**) The uptake of guest molecules from the liquid phase can be investigated by the same sample by connecting the QCM cell to a peristaltic pump, so that either the pure solvent or the guest-molecule solution can be pumped through the QCM cell. Switching from the pure gas flow, or pure solvent, to the enriched gas flow, or solution, results in the uptake of the guest molecules by the surface-mounted metal-organic framework (SURMOF). The mass changes are recorded by QCM.

Mass changes of the sample on the QCM sensor can be determined by the Sauerbrey equation [30], $\Delta m = C \times \Delta f_n/n$, with Δm denoting the mass change, C the Sauerbrey constant (here: 17.7 $ng \cdot cm^{-2} \cdot Hz^{-1}$) and n the overtone number. The sample thickness was determined from the change of the resonance frequency f_1 of the QCM sensor before (4,952,150 Hz) and after (4,951,556 Hz) the SURMOF synthesis. A SURMOF mass of 10.5 μg·cm^{-2} follows, which corresponds to a SURMOF thickness of about 100 nm.

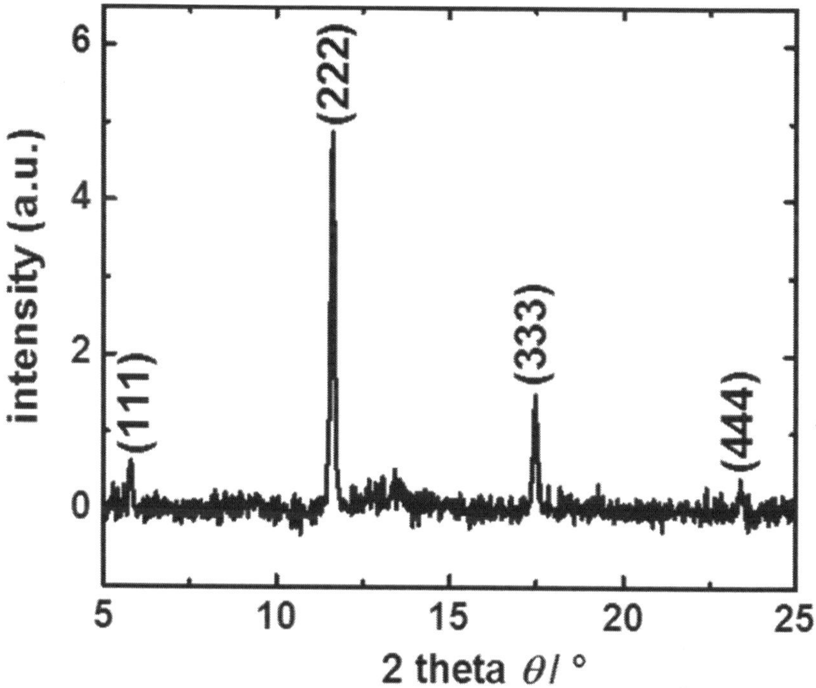

Figure 2. X-ray diffractogram of the HKUST-1 SURMOF. The diffraction reflexes at 5.8° (111), 11.6° (222), 17.5° (333) and 23.3° (444) demonstrate the crystallinity of the HKUST-1 SURMOF as well as the oriented growth in [111] direction. The X-ray diffractogram was measured with X-ray radiation of 0.154 nm wavelength by using an out-of-plane X-ray diffractometer of type Bruker D8 Advance.

The uptake of ferrocene from the gas phase is performed by connecting the QCM cell to a gas flow system [26]. The flow of the pure carrier gas (argon with a flow rate of 10 mL·min^{-1}) is instantly switched to a flow passing over the compound of the guest molecules (here: ferrocene powder with a saturation vapor pressure of about 1

Pa at room temperature [31]) resulting in a gas flow enriched with the vapor of the guest molecules. When the enriched gas flow passes through the QCM cell, the guest molecules are diffusing into the SURMOF and the uptake can be studied by the mass change of the SURMOF on the QCM sensor. It should be noted that the concentration of argon in MOFs of type HKUST-1 at room temperature and a pressure of about 1 bar is very small [32]. Therefore, the argon molecules in the pores as well as the diffusion of argon during the uptake of the guest molecules can be neglected and the activated MOF pores can be considered as empty.

The uptake from the liquid phase is investigated by initially pumping the pure solvent (*i.e.*, ethanol or *n*-hexane) through the QCM cell. Then, the solvent is instantly switched to the respective ferrocene solution. Here, ferrocene concentrations were chosen which correspond to 90% of the saturation solubilities [33], *i.e.*, 90 mM ferrocene in ethanol and 155 mM ferrocene in *n*-hexane. To ensure that the uptake by the SURMOF is controlled by diffusion in the nanopores and not by convection through the QCM cell, the flow rates of the gases and liquids through the QCM cell need to be large enough Here, it was set 10 $mL \cdot min^{-1}$ for the gas flow and roughly 1 $mL \cdot min^{-1}$ for the liquid phase.

To increase the significance of the data and to exclude artifacts, the uptake of the guest molecules from the liquid and the gas phase are alternatively repeated. The uptake curves of the ferrocene by the SURMOF are shown in Figure 3. All experiments were performed at a temperature of 25 °C. Before each experiment, the sample was activated in a flow of pure argon at 65 °C for several hours, to guarantee that no molecules, such as water, were adsorbed in the pores and to guarantee reproducible results.

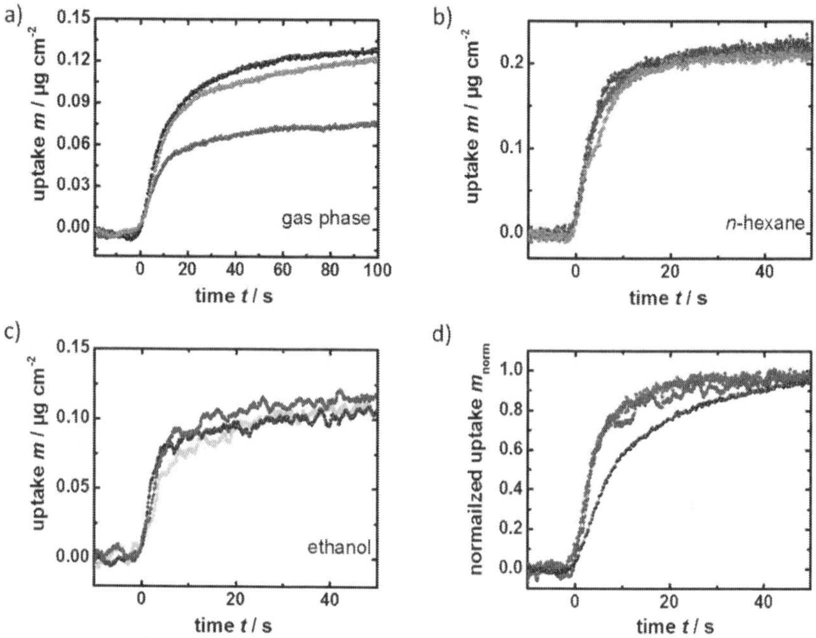

Figure 3. Ferrocene uptake by HKUST-1. The uptake curves of ferrocene vapor in argon carrier gas (**a**); of ferrocene in *n*-hexane (**b**); and of ferrocene in ethanol (**c**); are measured by QCM. For comparison of the uptake kinetics, the normalized uptake by the SURMOF from the liquid and the gas phase are shown in (**d**). The black sphere show the uptake of the ferrocene vapor, red of the *n*-hexanic ferrocene solution and blue of the ethanolic ferrocene solution.

3. RESULTS AND DISCUSSIONS

The uptake curves recorded by QCM enable an analysis of the equilibrium concentration and of the mass transfer parameter, *i.e.*, the diffusion coefficient. Since the SURMOF pores which are activated in argon can be considered as empty, the QCM experiments allow a straightforward quantification of the uptake amount, as demonstrated in references [34,35,36]. The ferrocene uptake is determined to be 0.11 ± 0.02 $\mu g \cdot cm^{-2}$, which corresponds to roughly

3/4 ferrocene compounds per HKUST-1 unit cell. Since ferrocene is a solid which slowly sublimates, it can be assumed that the obtained partial pressure of ferrocene is (significantly) smaller than the saturated vapor pressure of ferrocene, resulting in an equilibrium concentration which is (much) smaller than the maximum ferrocene concentration in the MOF pores.

The uptake from the liquid phase is more complex. The pores are initially filled with the solvent. When the investigated guest molecules are diffusing into the pores, the solvent molecules are displaced. Therefore, the determination of the uptake amount is not straightforward and requires the knowledge of the densities of the solvent in the MOF pores, with and without ferrocene. Due to the fact that the mass change is proportional to the ferrocene uptake, the recorded uptake curves can be used to investigate the kinetic properties and to determine the diffusion coefficient of the ferrocene compound in solution.

It was determined from the uptake curves (Figure 3) that the ferrocene uptake from the gas phase proceeds with an average time constant τ of 13 ± 4 s, from the *n*-hexanic solution with a time constant of 4.9 ± 1.1 s and from the ethanolic solutions with a time constant of 5.8 ± 1.9 s. The error denotes the standard deviation of the experimental results. By using the relationship $D = l^2/(3\tau)$ (reference [37]) and using the determined SURMOF thickness l of about 100 nm, the diffusion coefficient D for ferrocene in the (initially) empty MOF, *i.e.*, of ferrocene during the uptake from the gas phase, was determined to be (2.5 ± 0.7) × 10^{-16}m^2·s^{-1}. For the uptake from the liquid phase, ferrocene diffusion coefficients of (6.8 ± 1.6) × 10^{-16} m^2·s^{-1} in the *n*-hexanic solution and (5.6 ± 1.8) × 10^{-16} m^2·s^{-1} in the ethanolic solution were calculated. These experimental findings were reproduced by an HKUST-1 SURMOF of 110 nm thickness (sample 2), where diffusion coefficients of (1.6 ± 0.5) × 10^{-16} m^2·s^{-1} from the gas phase, (6.0 ± 0.5) × 10^{-16} m^2·s^{-1} from the hexanic solution and (7.5 ± 3.1) × 10^{-16} m^2·s^{-1} from the ethanolic solution were determined, see Figure 4.

This means that the uptake from the ferrocene solution by the solvent-filled HKUST-1 proceeds roughly 3-times faster than the uptake of the ferrocene vapor by the empty HKUST-1 framework. This is remarkable, since the solvent molecules have to leave the pores to allow the ferrocene to enter the solvent-filled MOF pores. It may be assumed that the fast diffusion in the solvent is caused by the interaction of the ethanol or*n*-hexane solvent molecules, which occupy attractive adsorption sites in the MOF framework. Thus, the ferrocene molecules are less strongly adsorbed in the pores and can diffuse faster through the framework. It can be assumed that the diffusion of small molecules like ethanol and *n*-hexane is much faster than of ferrocene. Therefore, the counter diffusion of the solvent molecules does not slow down the ferrocene uptake.

A similar behavior was found in reference [38], where the self-diffusion coefficient of methanol in MOFs of type ZIF-8 was found to be larger than the transport diffusion coefficient. The reason for this anomaly is the strong interaction between the methanol molecules and the "intercage hopping" diffusion process in the ZIF-8 framework.

The pore windows of HKUST-1 have a diameter of roughly 0.9 nm, which is significantly larger than ferrocene with a kinetic diameter of about 0.66 nm [39]. In a previous study, the uptake of ferrocene by MOFs of type $Cu_2(ndc)_2(dabco)$ with pore window diameters of 0.57 nm was investigated [26]. There, the diffusion coefficient of ferrocene is three orders of magnitude smaller than in HKUST-1. This shows that the pore size has a dramatic impact of the diffusion properties, especially if the pore diameter and the guest molecules have a comparable size.

It should be noted that, due to the fact that the pristine samples were used directly after the synthesis, surface barriers [40,41,42], which may hinder the uptake and release of the guest molecules, are neglected during the analysis of the data [25].

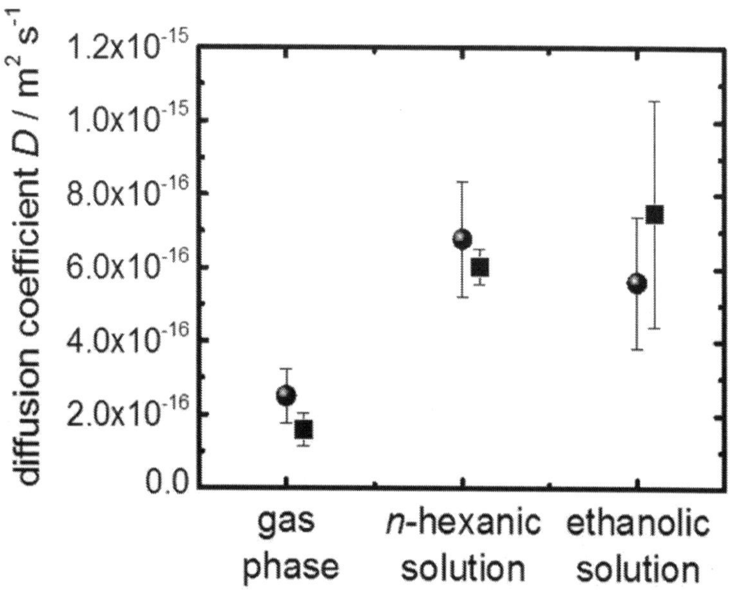

Figure 4. Diffusion coefficients of ferrocene in HKUST-1. The average diffusion coefficients with the respective standard deviation are shown for the ferrocene vapor in the argon carrier gas, for ferrocene in *n*-hexanic solution and for ferrocene in ethanolic solution. The results of both samples are shown; spheres: sample 1, squares: sample 2.

4. CONCLUSIONS

The diffusion of ferrocene in MOFs of type HKUST-1 was investigated by using thin MOF films in combination with QCM. For the first time, the diffusion of the guest molecules in the gas phase, where the guest molecules are diffusing into an initially empty framework, and in the liquid phase, where the guest molecules are diffusing into a solvent-filled framework, are compared. It was found that in the MOF, the diffusion coefficient of ferrocene in the liquid- and in the gas-phase are in the same order of magnitude; with a ferrocene diffusion coefficient which is roughly 3-times smaller in the empty MOF (*i.e.*, uptake from the gas phase) than in the liquid-filled MOF (*i.e.*, uptake from the ferrocene solutions). It may be assumed that other liquids may have a larger impact on the diffusion and significantly slow down or accelerate the ferrocene uptake.

Due to their general applicability, SURMOF films in combination with QCM show great potential for investigating the diffusion of guest molecules in MOFs. Not only the uptake from the gas and from the liquid phase can be investigated, but also guest-host systems with diffusion coefficients ranging over at least six orders of magnitude can be studied. So far, mass transfers with diffusion coefficients in the range of $6 \times 10^{-19}\ m^2 \cdot s^{-1}$ [26] to $6 \times 10^{-13}\ m^2 \cdot s^{-1}$ [25] have been investigated. It might be assumed that even slower or faster diffusion processes can be investigated, for instance by fully utilizing the time resolution and by increasing or decreasing the SURMOF thickness.

ACKNOWLEDGMENTS

W.Z. is grateful for a PhD fellowship donated by the China Scholarship Council (CSC). L.H. is indebted to the Baden-Württemberg Stiftung for the financial support of this research project by the Eliteprogramme for Postdocs. We acknowledge support by *Deutsche Forschungsgemeinschaft* and Open Access Publishing Fund of Karlsruhe Institute of Technology.

AUTHOR CONTRIBUTIONS

All authors have approved the final version of the manuscript. W.Z. synthesized and characterized the samples and performed the experiments. L.H. planned and evaluated the experiments.

CONFLICTS OF INTEREST

The authors declare no competing financial interest.

REFERENCES

1. Ferey, G. Hybrid porous solids: past, present, future. *Chem. Soc. Rev.* **2008**, *37*, 191–214.
2. Sanchez, C.; Shea, K.J.; Kitagawa, S. Recent progress in hybrid materials science. *Chem. Soc. Rev.* **2011**,*40*, 471–472.
3. Kreno, L.E.; Leong, K.; Farha, O.K.; Allendorf, M.; Van Duyne, R.P.; Hupp, J.T. Metal-Organic Framework Materials as Chemical Sensors. *Chem. Rev.* **2011**, *112*, 1105–1125.
4. Furukawa, H.; Cordova, K.E.; O'Keeffe, M.; Yaghi, O.M. The Chemistry and Applications of Metal-Organic Frameworks. *Science* **2013**, *341*, 1230444.
5. Caro, J. Are MOF membranes better in gas separation than those made of zeolites? *Curr. Opin. Chem. Eng.* **2011**, *1*, 77–83.
6. Kärger, J.; Binder, T.; Chmelik, C.; Hibbe, F.; Krautscheid, H.; Krishna, R.; Weitkamp, J. Microimaging of transient guest profiles to monitor mass transfer in nanoporous materials. *Nat. Mater.* **2014**, *13*, 333–343.
7. Chmelik, C.; Heinke, L.; Kortunov, P.; Li, J.; Olson, D.; Tzoulaki, D.; Weitkamp, J.; Kärger, J. Ensemble Measurement of Diffusion: Novel Beauty and Evidence. *Chem. Phys. Chem.* **2009**, *10*, 2623–2627.
8. Nobar, S.N.; Farooq, S. Experimental and modeling study of adsorption and diffusion of gases in Cu-BTC. *Chem. Eng. Sci.* **2012**, *84*, 801–813.

9. Li, J.R.; Kuppler, R.J.; Zhou, H.C. Selective gas adsorption and separation in metal-organic frameworks. *Chem. Soc. Rev.* **2009**, *38*, 1477–1504.

10. Skoulidas, A.I.; Sholl, D.S. Self-diffusion and transport diffusion of light gases in metal-organic framework materials assessed using molecular dynamics simulations. *J. Phys. Chem. B* **2005**, *109*, 15760–15768.

11. Stallmach, F.; Groger, S.; Kunzel, V.; Kärger, J.; Yaghi, O.M.; Hesse, M.; Müller, U. NMR studies on the diffusion of hydrocarbons on the metal-organic framework material MOF-5. *Angew. Chem. Int. Ed.***2006**, *45*, 2123–2126.

12. Wang, C.; Lin, W. Diffusion-Controlled Luminescence Quenching in Metal-Organic Frameworks. *J. Am. Chem. Soc.* **2011**, *133*, 4232–4235.

13. Garcia, E.R.; Medina, R.L.; Lozano, M.M.; Perez, I.H.; Valero, M.J.; Franco, A.M.M. Adsorption of Azo-Dye Orange II from Aqueous Solutions Using a Metal-Organic Framework Material: Iron-Benzenetricarboxylate. *Materials* **2014**, *7*, 8037–8057.

14. Shekhah, O.; Liu, J.; Fischer, R.A.; Wöll, C. MOF thin films: existing and future applications. *Chem. Soc. Rev.* **2011**, *40*, 1081–1106.

15. Shekhah, O.; Wang, H.; Kowarik, S.; Schreiber, F.; Paulus, M.; Tolan, M.; Sternemann, C.; Evers, F.; *et al.* Step-by-step route for the synthesis of metal-organic frameworks. *J. Am. Chem. Soc.* **2007**, *129*, 15118–15119.

16. Heinke, L.; Tu, M.; Wannapaiboon, S.; Fischer, R.A.; Wöll, C. Surface-mounted metal-organic frameworks for applications in sensing and separation. *Microporous Mesoporous Mater* **2015**.

17. Kind, M.; Wöll, C. Organic surfaces exposed by self-assembled organothiol monolayers: Preparation, characterization, and application. *Prog. Surf. Sci.* **2009**, *84*, 230–278.

18. Zacher, D.; Yusenko, K.; Betard, A.; Henke, S.; Molon, M.; Ladnorg, T.; Shekhah, O.; Schupbach, B.; *et al.* Liquid-Phase Epitaxy of Multicomponent Layer-Based Porous Coordination Polymer Thin Films of M(L)(P)0.5 Type: Importance of Deposition Sequence on the Oriented Growth. *Chem. A Eur. J.* **2011**,*17*, 1448–1455.

19. Munuera, C.; Shekhah, O.; Wang, H.; Wöll, C.; Ocal, C. The controlled growth of oriented metal-organic frameworks on functionalized surfaces as followed by scanning force microscopy. *Phys. Chem. Chem. Phys.* **2008**, *10*, 7257–7261.

20. Gu, Z.G.; Pfriem, A.; Hamsch, S.; Breitwieser, H.; Wohlgemuth, J.; Heinke, L.; Gliemann, H.; Wöll, C. Transparent films of metal-organic frameworks for optical applications. *Microporous Mesoporous Mater.* **2015**, *211*, 82–87.

21. Shekhah, O.; Wang, H.; Paradinas, M.; Ocal, C.; Schupbach, B.; Terfort, A.; Zacher, D.; Fischer, R.A.; *et al.* Controlling interpenetration in metal-organic frameworks by liquid-phase epitaxy. *Nat. Mater.***2009**, *8*, 481–484.

22. Wang, Z.; Liu, J.; Lukose, B.; Gu, Z.; Weidler, P.G.; Gliemann, H.; Heine, T.; Wöll, C. Nanoporous Designer Solids with Huge Lattice Constant Gradients: Multiheteroepitaxy of Metal-Organic Frameworks. *Nano Lett.* **2014**, *14*, 1526–1529.

23. Heinke, L.; Cakici, M.; Dommaschk, M.; Grosjean, S.; Herges, R.; Bräse, S.; Wöll, C. Photoswitching in Two-Component Surface-Mounted Metal-Organic Frameworks: Optically Triggered Release from a Molecular Container. *Acs Nano* **2014**, *8*, 1463–1467.

24. Zybaylo, O.; Shekhah, O.; Wang, H.; Tafipolsky, M.; Schmid, R.; Johannsmann, D.; Wöll, C. A novel method to measure diffusion coefficients in porous metal-organic frameworks. *Physical Chem. Chem. Phys.* **2010**, *12*, 8092–8097.

25. Heinke, L.; Gu, Z.; Wöll, C. The surface barrier phenomenon at the loading of metal-organic frameworks. *Nat. Commun.* **2014**, *5*, 4562.

26. Heinke, L.; Wöll, C. Adsorption and diffusion in thin films of nanoporous metal-organic frameworks: Ferrocene in SURMOF Cu-2(ndc)(2)(dabco). *Phys. Chem. Chem. Phys.* **2013**, *15*, 9295–9299.

27. Stavila, V.; Volponi, J.; Katzenmeyer, A.M.; Dixon, M.C.; Allendorf, M.D. Kinetics and mechanism of metal-organic framework thin film growth: systematic investigation of HKUST-1 deposition on QCM electrodes. *Chem. Sci.* **2012**, *3*, 1531–1540.

28. Chui, S.S.Y.; Lo, S.M.F.; Charmant, J.P.H.; Orpen, A.G.; Williams, I.D. A chemically functionalizable nanoporous material Cu-3(TMA)(2)(H2O)(3) (n). *Science* **1999**, *283*, 1148–1150.

29. Arslan, H.K.; Shekhah, O.; Wohlgemuth, J.; Franzreb, M.; Fischer, R.A.; Wöll, C. High-Throughput Fabrication of Uniform and Homogenous MOF Coatings. *Adv. Funct. Mater.* **2011**, *21*, 4228–4231.

30. Sauerbrey, G. Verwendung von Schwingquarzen zur Wägung dünner Schichten und zur Mikrowägung. *Z. Fur. Phys.* **1959**, *155*, 206–222.

31. Monte, M.J.S.; Santos, L.M.B.F.; Fulem, M.; Fonseca, J.M.S.; Sousa, C.A.D. New static apparatus and vapor pressure of reference materials: Naphthalene, benzoic acid, benzophenone, and ferrocene. *J. Chem. Eng. Data* **2006**, *51*, 757–766.

32. Krungleviciute, V.; Lask, K.; Heroux, L.; Migone, A.D.; Lee, J.Y.; Li, J.; Skoulidas, A. Argon Adsorption on Cu_3(Benzene-1,3,5-tricarboxylate)$_2$(H_2O)$_3$ Metal–Organic Framework. *Langmuir* **2007**,*23*, 3106–3109.

33. De Fina, K.M.; Ezell, C.; Acree, W.E. Solubility of ferrocene in organic nonelectrolyte solvents. Comparison of observed versus predicted values based upon mobile order theory. *Phys. Chem. Liq.***2001**, *39*, 699–710.

34. Cakici, M.; Gu, Z.G.; Nieger, M.; Burck, J.; Heinke, L.; Bräse, S. Planar-chiral building blocks for metal-organic frameworks. *Chem. Commun.* **2015**, *51*, 4796–4798.

35. Gu, Z.; Grosjean, S.; Bräse, S.; Wöll, C.; Heinke, L. Enantioselective Adsorption in Homochiral Metal-Organic Frameworks: The Pore Size Influence. *Chem. Commun.* **2015**, *51*, 8998–9001.

36. Wang, Z.; Heinke, L.; Jelic, J.; Cakici, M.; Dommaschk, M.; Maurer, R.J.; Oberhofer, H.; Grosjean, S.; *et al.* Photoswitching in nanoporous, crystalline solids: An experimental and theoretical study for azobenzene linkers incorporated in metal-organic frameworks. *Phys. Chem. Chem. Phys.* **2015**, *17*, 14582–14587.

37. Kärger, J.; Ruthven, D.M.; Theodorou, D.N. *Diffusion in Nanoporous Materials*; Wiley-VCH: Weinheim, Germany, 2012.

38. Chmelik, C.; Bux, H.; Caro, J.; Heinke, L.; Hibbe, F.; Titze, T.; Kärger, J. Mass Transfer in a Nanoscale Material Enhanced by an Opposing Flux. *Phys. Rev. Lett.* **2010**, *104*, 085902.

39. Chidsey, C.E.D.; Bertozzi, C.R.; Putvinski, T.M.; Mujsce, A.M. Coadsorption of ferrocene-terminated and unsubstituted alkanethiols on gold: Electroactive self-assembled monolayers. *J. Am. Chem. Soc.***1990**, *112*, 4301–4306.

40. Heinke, L.; Kärger, J. Correlating Surface Permeability with Intracrystalline Diffusivity in Nanoporous Solids. *Phys. Rev. Lett.* **2011**, *106*, 074501.

41. Kärger, J. In-depth study of surface resistances in nanoporous materials by microscopic diffusion measurement. *Microporous Mesoporous Mater.* **2014**, *189*, 126–135.

42. Heinke, L.; Kortunov, P.; Tzoulaki, D.; Kärger, J. Exchange dynamics at the interface of nanoporous materials with their surroundings. *Phys. Rev. Lett.* **2007**, *99*, 228301.

CHAPTER 4

Two-Dimensional Simulation of Mass Transfer in Unitized Regenerative Fuel Cells under Operation Mode Switching

Lulu Wang [1], Hang Guo [1,2,*], Fang Ye [1] and Chongfang Ma [1]

[1]Key Laboratory of Enhanced Heat Transfer and Energy Conservation, Ministry of Education and Beijing Key Laboratory of Heat Transfer and Energy Conversion, College of Environmental and Energy Engineering, Beijing University of Technology, Beijing 100124, China

[2]Collaborative Innovation Center of Electric Vehicles in Beijing, Beijing 100081, China

ABSTRACT

A two-dimensional, single-phase, isothermal, multicomponent, transient model is built to investigate the transport phenomena in unitized regenerative fuel cells (URFCs) under the condition of switching from the fuel cell (FC) mode to the water electrolysis (WE) mode. The model is coupled with an electrochemical reaction. The proton exchange membrane (PEM) is selected as the solid electrolyte of the URFC. The work is motivated by the need to elucidate the complex mass transfer and electrochemical process under operation mode switching in order to improve the performance of PEM URFC. A set of governing equations, including conservation of mass, momentum, species, and charge, are considered. These equations are solved by the finite element method. The simulation results indicate the distributions of hydrogen, oxygen, water mass fraction, and electrolyte potential response to the transient phenomena via saltation under operation mode switching. The hydrogen mass fraction gradients are smaller than the oxygen mass fraction gradients. The average mass fractions of the reactants (oxygen and hydrogen) and product (water) exhibit evident differences between each layer in the steady state of the FC mode. By contrast, the average mass fractions of the reactant (water) and products (oxygen and hydrogen) exhibit only slight differences between each layer in the steady state of the WE mode. Under either the FC mode or the WE mode, the duration of the transient state is only approximately 0.2 s.

KEYWORDS

numerical simulation; regenerative fuel cell; operation mode switching; transport phenomenon; two-dimensional

1. INTRODUCTION

URFCs based on PEM electrolyte are reversible electrochemical devices capable of operating in water electrolysis (WE) mode and H_2/O_2 FC mode. In the WE mode, water molecules are split into hydrogen and oxygen with the assistance of external voltage. In the FC mode, hydrogen and oxygen molecules are combined to form water and generate electricity [1,2,3,4]. URFCs are used extensively, not only for space or military applications, but also for solar-powered aircraft, satellites, and micro-spacecraft applications [5,6,7]. Proton exchange membrane (PEM) is a convenient electrolyte technology for both WEs and FCs, as it allows for low operational temperature, quick start, fast response, high power and energy densities [8].

The vast literature on experimental research and modeling of PEMFC deal with the water, heat management (e.g., [9,10,11]) and mass transport (e.g., [12,13,14]). Hu *et al.* [13] built a comprehensive model which accounted for the transport phenomena occurring in the PEMFC such as mass transfer, electrochemical kinetics and charge transport. Singh *et al.* [14] developed a theoretical model to simulate transport phenomena in a PEMFC. The work was dedicated to understanding the transport processes in fuel cells for the purpose of improving heat and water management, and alleviating mass transport limitations. Several articles have been published on experimental research and modeling of PEMWE relating to the hydrogen storage, flow and heat transfer (e.g., [15,16,17,18]). Grigoriev *et al.* [18] built a model to identify the important physical and chemical phenomena at high pressure in PEM water electrolysers. The model was also used to optimize the performance of a PEM cell. A large amount of literature on experimental PEM URFC research focuses mainly on the preparation techniques for membrane electrode assembly (MEA), the selection of gas diffusion layer (GDL) materials, and the treatment of corrosion-resistant GDL surfaces in terms of content. In addition, system optimization, the effects of structural and operating parameters on the performance of URFCs have also been experimentally investigated [19,20,21,22,23,24]. The literature on modeling of PEM URFC is extremely limited and very few articles have been published. Guarnieri *et al.* [8] presented a performance model of a PEM URFC via providing voltage, power, and efficiency under varying load conditions as functions of the controlling physical quantities: temperature, pressure, concentration, and humidification. The model was used as a tool for investigating optimized cell/stack designs and operational conditions. Doddathimmaiah and Andrews [25] constructed a computer model based on Excel and Visual Basic to generate voltage-current curves in both WE and FC modes for PEM URFCs with a range of membrane electrode assembly (MEA) characteristics by modifying the standard Butler-Volmer equation. The effect of key factors such as exchange current densities and charge transfer coefficient on cell performance was analyzed. The two models of PEM URFC mentioned above are both steady state, few models can be found for transient performance investigations.

URFCs play a significant role in the promotion of renewable energy sources (wind or solar energy). In practical applications, URFCs can combine with wind or solar energy to store hydrogen and produce electricity [26]. Therefore, switching between both modes is required to achieve this. URFCs involve several problems in the process of operation mode switching, including material stability, corrosion of carbon-base materials (carbon used as catalyst carrier, gas diffusion layer, and bipolar plate) at the anode during WE mode. These problems result in the reduction of catalyst activity and mass transport limitations. Hence, the performance degrades rapidly [4]. Meanwhile, there are complicated transient interactions between electrochemical reactions and transport processes during the operation mode switching [27]. Jin *et al.* [27] built an isothermal two-dimensional (2D) transient model for regenerative solid oxide fuel cells to investigate complicated multi-physics process during the process of mode switching. Simulation results indicated the trend of internal parameter distributions, such as H_2/O_2/H_2O and ionic, electronic potentials. Similarly, it should be emphasized that an investigation into the fundamental mechanism under operation mode switching can offer important guidance in structure optimization, appropriate selection of operating parameters, and mass transfer enhancement for PEM URFC. However, no accessible literature can be found for this purpose.

This study investigates the transient behavior of URFCs under operation mode switching with a numerical simulation technique. The basic idea of modeling is as follows: (1) a two-dimensional (2D) model is selected because the study aims to observe the transport phenomena along the gas flow channel and the vertical location of the PEM; (2) water is maintained in the gaseous state to simplify the model into single-phase [28,29]; (3) the change in the internal temperature is not considered because the main objective of this study is to investigate the mass and electric transport phenomena; (4) the model involves multicomponent species; (5) operation mode switching is a transient process. (6) the model is coupled with an electrochemical reaction, which involves the electric transport between electrodes and chemicals. On the basis of the basic idea of modeling, a 2D, single-phase, isothermal, multicomponent, transient model is built to investigate the transport phenomena in PEM URFCs when switching from the FC mode to the WE mode. The model is coupled with an electrochemical reaction and considers a set of governing equations, including the conservation of mass, momentum, species, and charge. This study aims to elucidate the complex mass transfer and electrochemical process under operation mode switching.

2. MODEL DESCRIPTION

Figure 1 shows the computational domain of a 2D URFC model, which includes a gas flow channel, GDL, catalyst layer (CL) at the hydrogen and oxygen sides respectively, and a PEM sandwiched between the two sides. The gases in the gas flow channel at the hydrogen side are a mixture of H_2 and H_2O, and those in the gas flow channel at the oxygen side are a mixture of O_2 and H_2O.

2.1. Main Hypotheses of the Model

A 2D, single-phase, isothermal, multicomponent, transient model coupled with an electrochemical reaction is built in this study. The model makes the following assumptions:

1 The temperature inside the URFC is uniform at a constant value of 353 K. Any heat exchange is not considered [14,28].

2 The Reynolds number and velocity is low. Thus, the flow condition is laminar [13].

3 GDLs and CLs are isotropic porous media [13,28,29].

4 The PEM is impermeable to gas species [13].

5 Water is maintained in the gaseous state [28,29].

6 Gases are incompressible [14,28,29].

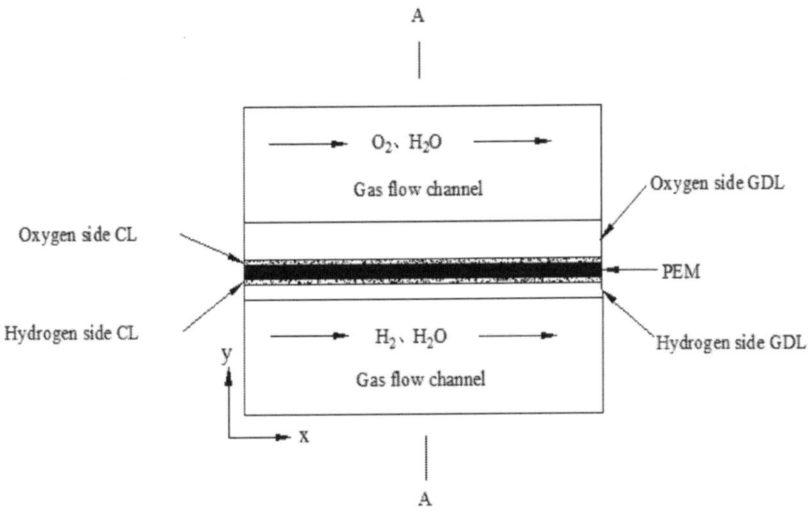

Figure 1. Computational domain of a 2D PEM URFC model.

2.2. Governing Equations

2.2.1. Charge Balance

Electrons and ions are transported between oxygen and hydrogen electrode, while PEM only allows ions to migrate through. The charge balance equation based on the generalized Ohm's law can be expressed as follows:

$$\nabla\left(-\sigma_l \nabla \phi_l\right) = \pm a_v i_{\mathrm{loc}} \tag{1}$$

$$\nabla\left(-\sigma_s \nabla \phi_s\right) = \pm a_v i_{\mathrm{loc}} \tag{2}$$

where σl and σs are ionic and electronic conductivities, respectively. ϕl and ϕs are ionic and electronic potentials, respectively. av is the active specific surface area. The signs in the right side of Equations (1) and (2) are dependent on the cell mode. Table 1 is the selection of signs for equations in FC and WE modes.

iloc in Equations (1) and (2) is the current density that can be described with the Butler-Volmer equation:

$$i_{\mathrm{loc}} = i_0 \left(\exp\left(\frac{\alpha_a F \eta}{RT}\right) - \exp\left(\frac{-\alpha_c F \eta}{RT}\right) \right) \tag{3}$$

where i0 is the exchange current density, which is dependent on reactant and production concentrations. αaand αc are the transfer coefficients for anode and cathode respectively. Overpotential η, is represented by the following equation:

$$\eta = \Phi_s - \Phi_l - E_{\mathrm{eq}} \tag{4}$$

$$E_{\mathrm{eq}} = E_{\mathrm{eq.ref}} + \mathrm{d}E_{\mathrm{eq}}/\mathrm{d}T\,(T - T_{\mathrm{ref}}) \tag{5}$$

where Eeq.ref is the reference equilibrium potential, Eeq is the equilibrium potential, which is a constant in the model because the temperature is constant.

Table 1. Selection of signs for equations in FC and WE modes.

Equations	Mode			
	FC Mode		WE Mode	
	Hydrogen Electrode	Oxygen Electrode	Hydrogen Electrode	Oxygen Electrode
$\nabla\left(-\sigma_l \nabla \phi_l\right) = \pm a_v i_{\mathrm{loc}}$	+	-	-	+
$\nabla\left(-\sigma_s \nabla \phi_s\right) = \pm a_v i_{\mathrm{loc}}$	-	+	+	-

2.2.2. Multicomponent Mass Transport

The gas transport is described with the Maxwell-Stefan's convection and diffusion equations as follows:

$$\rho\frac{\partial\omega_i}{\partial t} + \nabla\cdot\mathbf{j}_i + \rho\left(\mathbf{u}\cdot\nabla\right)\omega_i = R_i \tag{6}$$

$$R_i = \nu_i\frac{a_v i_{\text{loc}}}{n_i F} \tag{7}$$

$$\mathbf{j}_i = -\rho\omega_i\sum_k D_{ik}\mathbf{d}_k \tag{8}$$

$$\mathbf{d}_k = \nabla x_k + \left(x_k - \omega_k\right)\frac{\nabla p}{p} \tag{9}$$

$$x_k = \frac{\omega_k}{M_k}M \tag{10}$$

$$\frac{1}{M} = \sum_i\frac{\omega_i}{M_i} \tag{11}$$

Finally, Maxwell-Stefan's equation is:

$$\rho\frac{\partial\omega_i}{\partial t} + \nabla\left(\omega_i\rho\mathbf{u} - \rho\omega_i\sum_{k-1}^{Q}D_{ik}\left(\frac{M}{M_k}\left(\nabla\omega_k + \omega_k\frac{\nabla M}{M}\right) + \left(x_k - \omega_k\right)\frac{\nabla p}{p}\right)\right) = R_i$$

(12)

where ωi is the mass fraction of species *i*, ji is the mass flux relative to the mass average velocity, Dik is the multicomponent Fick diffusivity, dk is the diffusional driving force acting on species *k*, xk is the mole fraction, M is the mean molar mass, and the source term Ri is the rate expression describing its production or consumption.

2.2.3. Gas Flow Equations

Navier—Stokes equations are used to govern the flows in the gas flow channels:

$$\rho\frac{\partial \mathbf{u}}{\partial t}+\rho\left(\mathbf{u}\cdot\nabla\right)\mathbf{u}=\nabla\cdot\left[-p\mathbf{I}+\mu\left(\nabla\mathbf{u}+\left(\nabla\mathbf{u}\right)^{T}\right)-\frac{2}{3}\mu\left(\nabla\mathbf{u}\right)\mathbf{I}\right]$$

(13)

$$\frac{\partial\rho}{\partial t}+\nabla\cdot\left(\rho\mathbf{u}\right)=0 \tag{14}$$

Flow in porous electrodes is described with the following Brinkman equations:

$$\frac{\rho}{\varepsilon}\left(\frac{\partial \mathbf{u}}{\partial t}+\left(\mathbf{u}\cdot\nabla\right)\frac{\mathbf{u}}{\varepsilon}\right)=\nabla\cdot\left[-p\mathbf{I}+\frac{\mu}{\varepsilon}\left(\nabla\mathbf{u}+\left(\nabla\mathbf{u}\right)^{T}\right)-\frac{2\mu}{3\varepsilon}\left(\nabla\mathbf{u}\right)\mathbf{I}\right]-\left(\frac{\mu}{\kappa}+\frac{s_m}{\varepsilon^2}\right)\mathbf{u}$$

(15)

$$\frac{\partial\left(\varepsilon\rho\right)}{\partial t}+\nabla\cdot\left(\rho\mathbf{u}\right)=s_m \tag{16}$$

where ε and κ, are the porosity and permeability of the gas diffusion layer or catalyst layer, respectively. In addition, the source term sm is closely associated with the current density:

sm=∑iavilocMiniF (17)

2.3. Initial and Boundary Conditions

The boundary conditions include the average velocity for the inlet of gas flow channel, exit pressure for the outlet of gas flow channel and no slip wall.

The initial conditions:

(1) For the GDL, CL at the H_2 side and PEM: $\phi l=0$, $\phi s=0$; for the GDL, CL at the O_2 side: $\phi l=0$,$\phi s=Vcell$ (the operating voltage);

(2) The initial values of the oxygen and hydrogen mass fractions are both 0.9.

3. ELEMENT INDEPENDENCE TEST AND MODEL VALIDATION

3.1. Element Independence Test

The governing equations are solved by the finite element method. The number of elements may have an influence on the results. Therefore, six different number of elements are generated for the model. Figure 2a,b show the oxygen and water mass fractions of a point located at the center of the gas flow channel at the oxygen side changing with the operating voltage under the six number of elements. Figure 2c shows the hydrogen mass fraction of a point located at the center of the gas flow channel at the hydrogen side changing with the operating voltage under the six number of elements. We can observe from Figure 2 that there are no changes in the oxygen, water and hydrogen mass fractions for the number larger than 1410. This observation indicates that the number of elements has no influence on the results. Thus, the number of elements selected is 1410 to reduce the computation time and enhance the accuracy of the results. Figure 3shows the mesh generation of the model. A structured quadrilateral mesh is created, and the number of the quadrilateral elements is 1410.

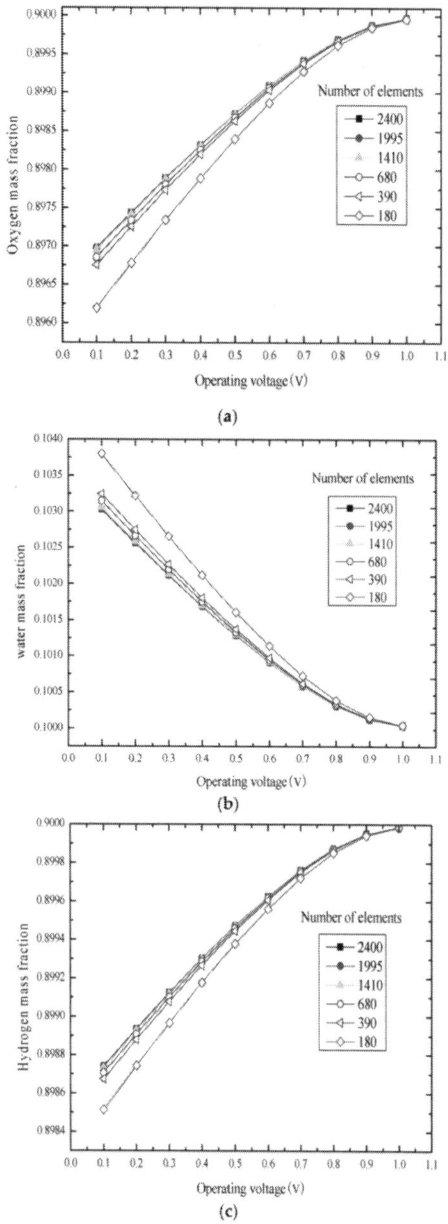

Figure 2. Element independence test for the present mode. (**a**) O_2 mass fraction; (**b**) O_2 side H_2O mass fraction; (**c**) H_2 mass fraction.

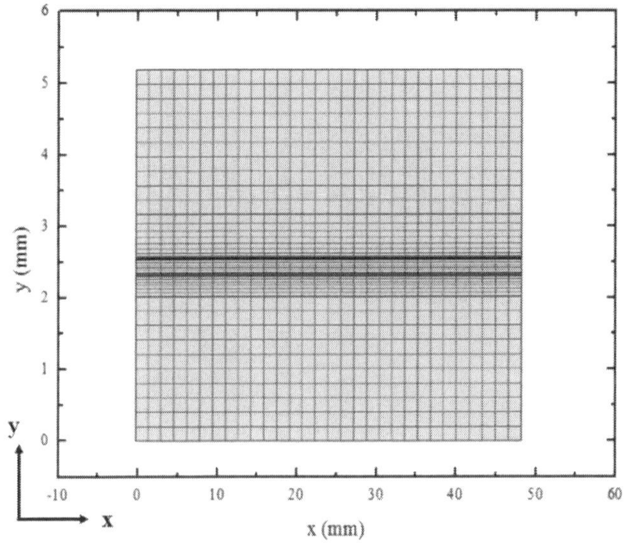

Figure 3. Mesh of the present model.

3.2. Model Validation

The results computed with this model at the steady state are compared with the experimental data in the literature [30], as shown in Figure 4. The numerical model and experimental data are analyzed under the same conditions (ambient temperature and pressure). Notably, the reference transfer current density is adjusted to achieve a good agreement between the present model and the experimental data. Figure 4 shows a slight difference in the I-V curves of the present model and the experimental data. The performance of the present model is better than that indicated by the experimental data in the FC mode. By contrast, the performance indicated by the experimental data is better than that of the present model in the WE mode. Such differences may be explained as follows. On the one hand, the details of the physical parameters used in the experiment are unknown, and the parameters used in the present model do not conform to the experiment fully. On the other hand, the model presents an assumption that water is maintained in the gaseous state.

The water generated in the experiment is in the liquid state and may degrade the cell performance by preventing the reactants from reaching the catalyst sites [29,31]. This phenomenon does not occur in the present model. Hence, the results in the experimental data are worse than those obtained with the present model. In the WE mode, water in the liquid state may be distributed more uniformly in the GDL compared with water in the gaseous state. In this case, the experimental data exhibit excellent performance. As such, we can conclude that the present model can be used for related simulations.

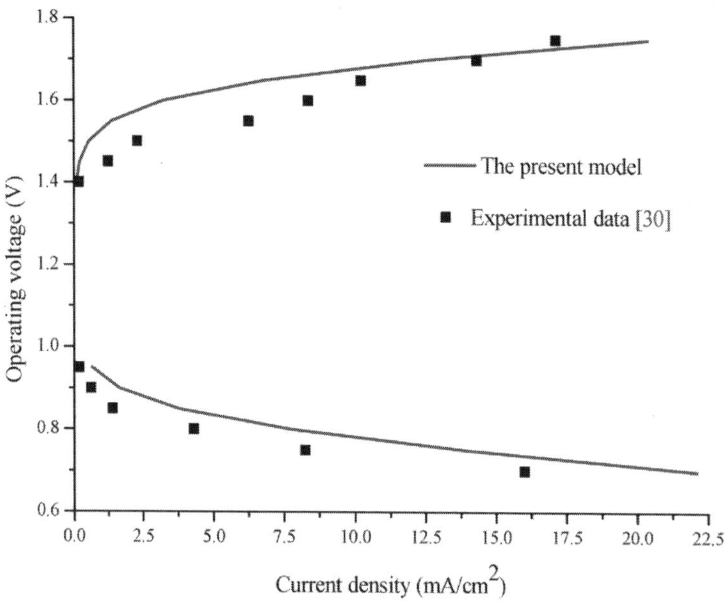

Figure 4. Comparison of the computed URFC performance with the experimental results.

4. RESULTS AND DISCUSSION

The transient response of the operating voltage to time under operation mode switching is outlined inFigure 5. The simulation time is set to 5 s. First, the cell functions in the FC mode with an operating voltage of 0.6 V, which is lower than the open circuit voltage (1.23 V). Then, at 0 s, the operating voltage changes from 0.6 V to 1.5 V, which is greater than the open circuit voltage. Correspondingly, the cell switches from the FC mode to the WE mode. Therefore, the minus and plus signs of time symbolize the cell in the FC and WE modes, respectively. The transient transport results under operation mode switching are eventually obtained via numerical simulation. The physical parameters of the URFC and the basic conditions used in this computation are listed in Table 2.

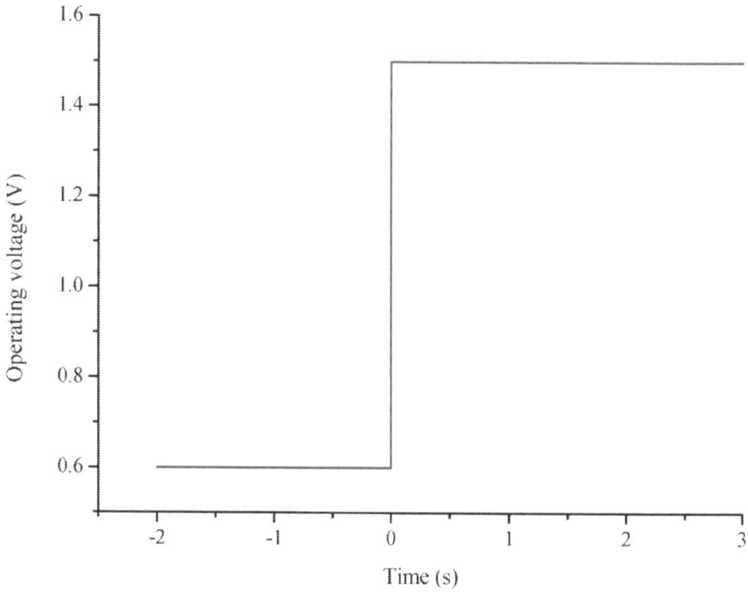

Figure 5. Transient response of the operating voltage to time under mode switching.

Figure 6 shows the hydrogen, oxygen, and water mass fraction distributions along line A–A (as shown inFigure 1, parallel to the y-axis at x = 0.024 m) at 0 s, which is a special time of the operation mode switching. The cell initially operates in the FC mode. Hydrogen and oxygen are transported from the gas flow channel to the GDL via convection and diffusion. Afterward, hydrogen and oxygen arrive at the CL via diffusion and are consumed at the hydrogen and oxygen electrodes, respectively. Notably, the hydrogen mass fraction decreases from the gas flow channel to the CL along line A–A at the hydrogen side. The oxygen mass fraction exhibits a similar trend in the gas flow channel to the CL along line A–A at the oxygen side. However, compared with the hydrogen and oxygen mass fractions, the water mass fraction at the oxygen side exhibits the opposite trend from the gas flow channel to the CL along line A–A at the oxygen side. This is due to the consumption of H_2, O_2 and the generation of H_2O during the electrochemical reaction.

Table 2. Physical parameters and basic conditions.

Length of channel/mm	48	Assumed
Gas flow channel width/mm	2	Assumed
Oxygen electrode GDL thickness/mm	0.6	Assumed
CL thickness/mm	0.028	Assumed
Hydrogen electrode GDL thickness/mm	0.3	Assumed
Membrane thickness/mm	0.178	Assumed
H_2 mass fraction	0.9	[32]
O_2 mass fraction	0.9	[32]
O_2/ H_2 inlet pressure/Pa	1.01×10^5	[32]
O_2 inlet velocity/$m \cdot s^{-1}$	1.18	[32]
H_2 inlet velocity/$m \cdot s^{-1}$	0.53	[32]
Oxygen/hydrogen electrode GDL permeability/m^2	1.18×10^{-11}	[32]
Membrane conductivity/$S \cdot m^{-1}$	1.4	[32]
Oxygen/hydrogen electrode GDL electrical conductivity/$S \cdot m^{-1}$	1000	[32]
H_2 reference concentration/$mol \cdot m^{-3}$	56.4	[33]
O_2 reference concentration/$mol \cdot m^{-3}$	40.8	[33]
Anodic transfer coefficient	0.5	[13]
Cathodic transfer coefficient	0.5	[13]
Operating temperature/K	353	[34]
Hydrogen electrode GDL porosity	0.4	[35]
Oxygen electrode GDL porosity	0.5	[36]
H_2/H_2O binary diffusion coefficient/$m^2 \cdot s^{-1}$	1.22×10^{-4}	Calculated
O_2/H_2O binary diffusion coefficient/$m^2 \cdot s^{-1}$	3.54×10^{-5}	Calculated
Oxygen/hydrogen electrode CL porosity	0.25	Assumed
Active specific surface area/m^{-1}	1.4×10^5	Assumed
Reference temperature/K	298.15	Assumed

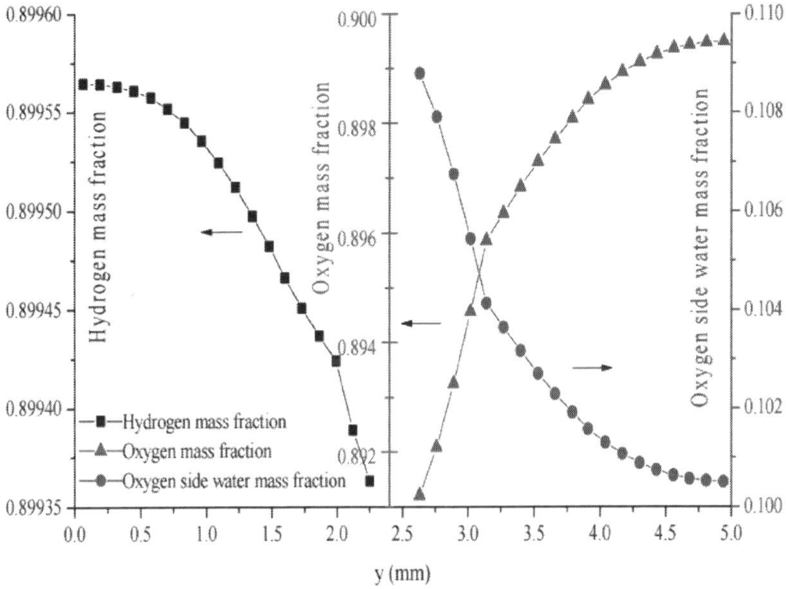

Figure 6. Parameter distribution along line A–A at 0 s.

Figure 7 shows the hydrogen, oxygen, and water mass fraction distributions at the oxygen side along line A–A at 3 s. At 0 s, the operating voltage changes suddenly from 0.6 V to 1.5 V. Thereafter, the switch from the FC mode to the WE mode is achieved. Therefore, the cell functions in the WE mode at 3 s. The water mass fraction at the oxygen side decreases from the gas flow channel to the CL along line A–A at the oxygen side. The minimum of water mass fraction is obtained at the CL. Nevertheless, the oxygen and hydrogen mass fractions exhibit opposite trends from the gas flow channel to the CL along line A–A at the oxygen and hydrogen sides, owing to the generation of O_2 and H_2 during the electrochemical reaction. The maximum mass fractions of oxygen and hydrogen are obtained at the CL.

Figure 8a,b show the time-dependent evolution of the average mass fractions of O_2 and H_2 in different layers. In the first −2 s, the average mass fractions of O_2 and H_2 decrease to the minimum in each layer at approximately 0.2 s and remain unchanged for the rest of time. The electrochemical reaction rate is rapid and only takes a short time to reach the steady state from the transient state in the FC mode. The average mass fractions of O_2 and H_2 exhibit a larger reduction in the CL than in the other two layers because O_2 and H_2 provided in the gas flow channel and supplied to the CL are consumed via the GDL at the oxygen and hydrogen sides, respectively. Differences in the minimum between each layer are evident in the FC mode. At 0 s to 3 s, the average mass fractions of O_2 and H_2 rapidly increase to the maximum from the minimum and maintain constant for the remaining time. The switch from the FC mode toward the WE mode is achieved at 0 s. The duration of the transient state is relatively short that the cell reaches the steady state in approximately 0.2 s. The average mass fractions of O_2 and H_2 exhibit a larger increase in the CL than in the other two layers because O_2 and H_2 are produced in the CL at the oxygen and hydrogen electrodes, respectively. A slight difference in the maximum mass fraction is observed between each layer in the WE mode. We conclude that the average mass fractions of O_2 and H_2 exhibit evident differences between each layer in the steady state of the FC mode and only slight differences between each layer in the steady state of the WE mode. The duration of the switch from the transient state to the steady state in the FC and WE modes is only approximately 0.2 s.

Figure 9 shows the time-dependent evolution of the average mass fractions of O_2 side H_2O in different layers. In the first −2 s, the average mass fractions of O_2 side H_2O increase to the maximum in approximately 0.2 s and maintain constant in different layers. The average mass fractions of O_2 side H_2O increase more significantly in the CL than in the other two layers because water is generated in the CL at the oxygen electrode. The differences in the maximum mass fraction of each layer are significant in the FC mode. At 0 s to 3 s, the average mass fractions of O_2 side H_2O rapidly decrease to the minimum from the maximum and remain unchanged for the rest of time. A significant reduction is observed in the CL compared with that in the other two layers because water is split in the CL at the oxygen electrode in the WE mode. A slight difference is observed in the minimum mass fraction between each layer.

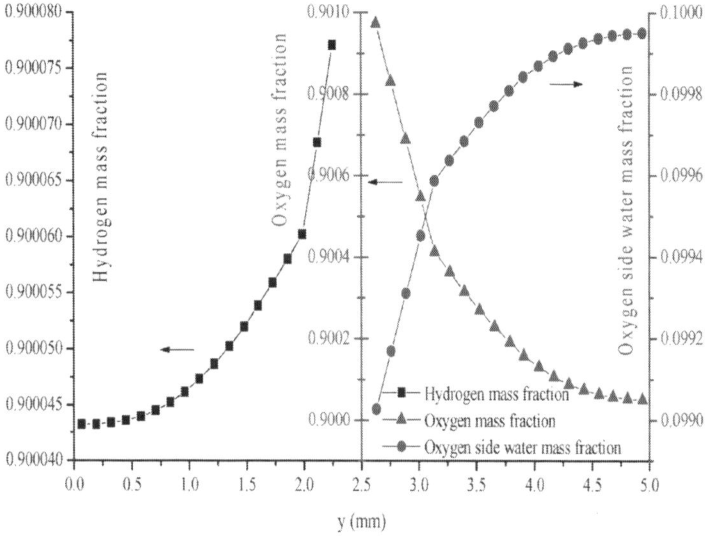

Figure 7. Parameter distributions along line A–A at 3 s.

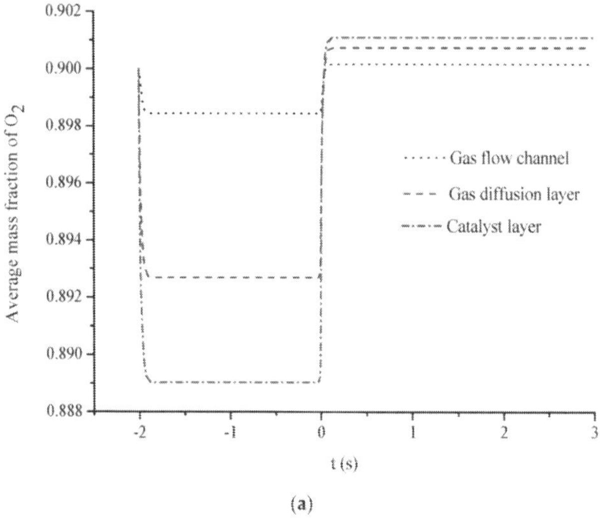

(a)

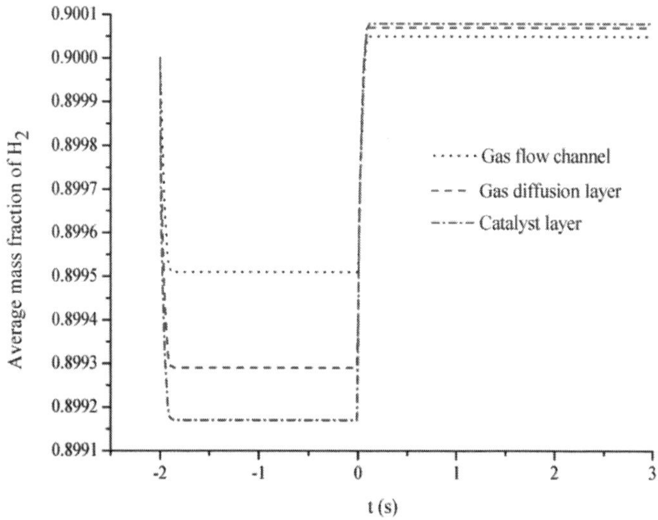

(**b**)

Figure 8. (**a**) Time-dependent evolution of O_2 mass fraction in different layers; (**b**) Time-dependent evolution of H_2 mass fraction in different layers.

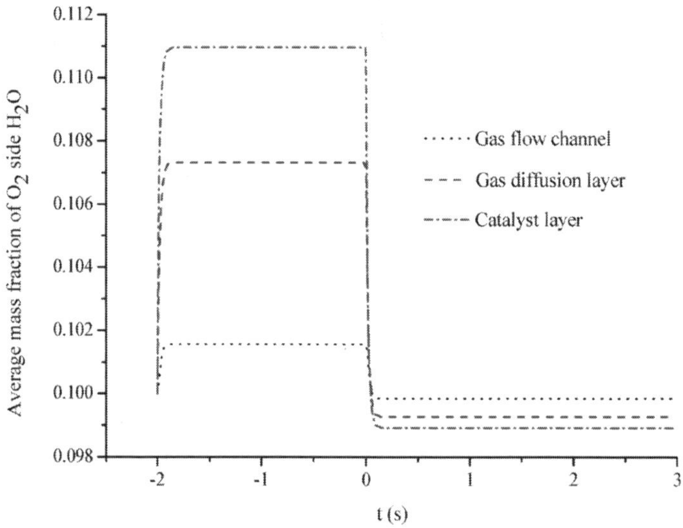

Figure 9. Time-dependent evolution of H_2O mass fraction in different layers.

Figure 10a,b show the 2D distributions of O_2, H_2, and H_2O mass fractions at −0.01 s before the operation mode switching. The cell is in the steady state at −0.01 s according to the Figure 8. Hydrogen mass fraction is distributed uniformly in each layer. On one hand, the electrochemical reaction consumes a little hydrogen, which results in a little reduction in the mass fraction. On the other hand, the excessive hydrogen is provided in the H_2 side gas flow channel, and is rapidly diffused on the surface of CL, a large amount of hydrogen is evenly distributed on the catalyst surface. The oxygen mass fraction decreases with an evident gradient from the gas flow channel to the CL at the oxygen side. However, the water mass fraction exhibits a different trend compared with hydrogen and oxygen; it increases distinctly from the gas flow channel to the CL. Figure 10c,d show the 2D distributions of O_2, H_2, and H_2O mass fractions at 0.01 s after the switching of the FC mode towards the WE mode. The cell is in the transient state at −0.01 s according to the Figure 8. The transient phenomena show that all values of the hydrogen and oxygen mass fractions increase, whereas the overall value of the water mass fraction decreases compared with that at −0.01 s. The hydrogen mass fraction decreases from the inlet of the gas flow channel to the outlet at the hydrogen side and slightly increases near the outlet compared with that at −0.01 s. The oxygen mass fraction also decreases with an evident gradient from the gas flow channel to the CL. Nevertheless, the water mass fraction exhibits the opposite trend along the same direction at the oxygen side. As such, we can conclude that the mass fractions of hydrogen, oxygen, and water respond to the sudden change of operating voltage via saltation under mode switching. The overall electrochemical reaction equation in FC mode is $2H_2 + O_2 \rightarrow 2H_2O$, we can find from this equation that it needs to consume 2 mol (4 g) H_2 and 1 mol (32 g) O_2 to generate 2 mol (36 g) H_2O, the consumption of H_2 mass is smaller than O_2 mass. Therefore, the hydrogen mass fraction gradients are smaller than the oxygen mass fraction gradients for the same scale legend.

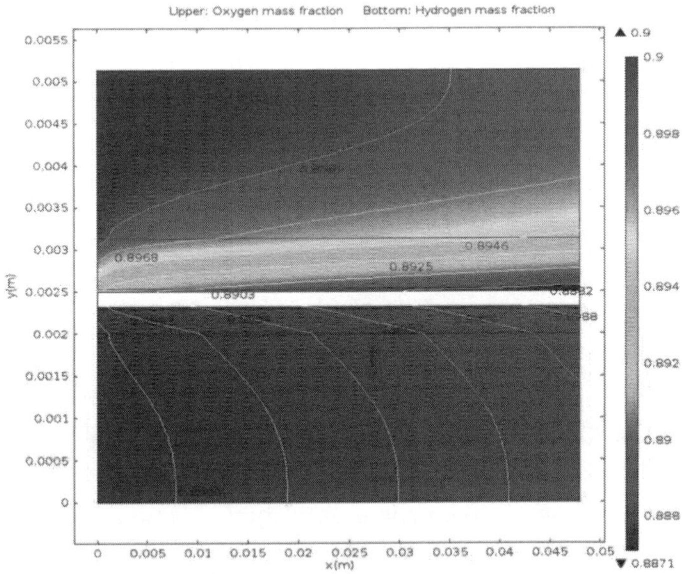
Upper: Oxygen mass fraction Bottom: Hydrogen mass fraction
0.0055
0.005
0.0045
0.004
0.0035
0.003
0.0025
0.002
0.0015
0.001
0.0005
0
y(m)
0 0.005 0.01 0.015 0.02 0.025 0.03 0.035 0.04 0.045 0.05
x(m)
▲ 0.9
0.9
0.898
0.896
0.894
0.892
0.89
0.888
▼ 0.8871
0.8968
0.8946
0.8925
0.8903

(a)

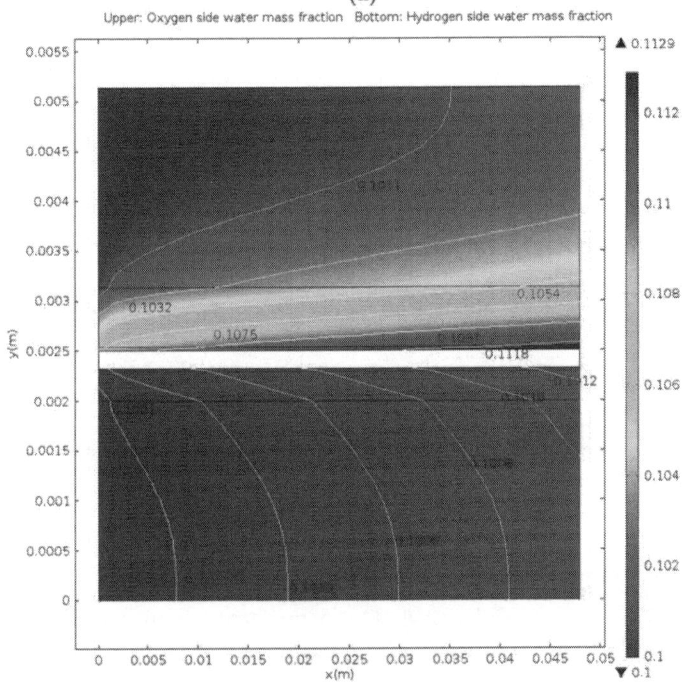
Upper: Oxygen side water mass fraction Bottom: Hydrogen side water mass fraction
0.0055
0.005
0.0045
0.004
0.0035
0.003
0.0025
0.002
0.0015
0.001
0.0005
0
y(m)
0 0.005 0.01 0.015 0.02 0.025 0.03 0.035 0.04 0.045 0.05
x(m)
▲ 0.1129
0.112
0.11
0.108
0.106
0.104
0.102
0.1
▼ 0.1
0.1032
0.1054
0.1075
0.1118

(b)

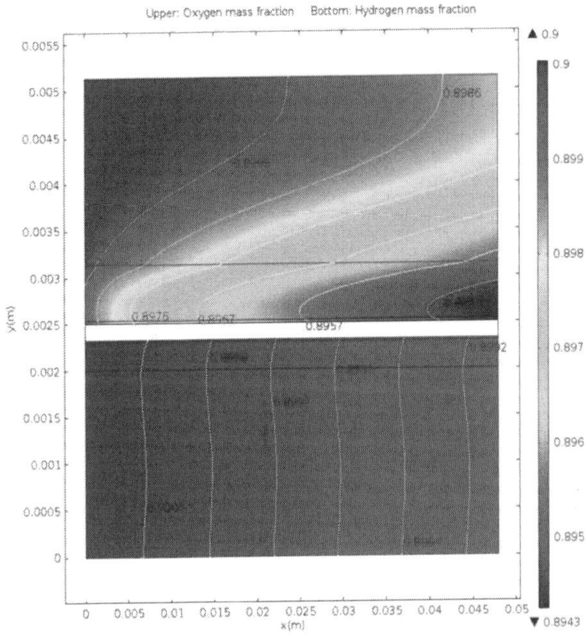

(c)

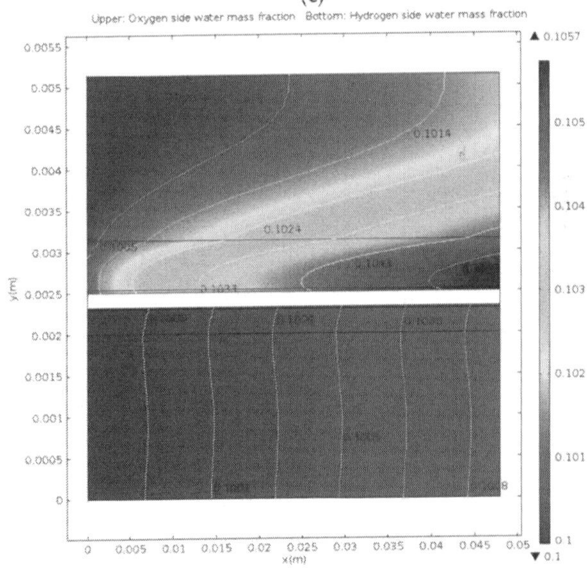

(d)

Figure 10. (**a**) Distributions of O_2 and H_2 mass fractions at −0.01 s; (**b**) Distributions of H_2O mass fractions at −0.01 s; (**c**) Distributions of $O_2$and H_2 mass fractions at 0.01 s; (**d**) Distributions of H_2O mass fractions at 0.01 s.

The evolution of the electronic potential with time along line A–A under mode switching is shown inFigure 11. The electronic potential is maintained at zero for the hydrogen electrode. However, the electronic potential changes from approximately 0.6 V in the FC mode to 1.5 V in the WE mode at the oxygen electrode during the transient process of mode switching.

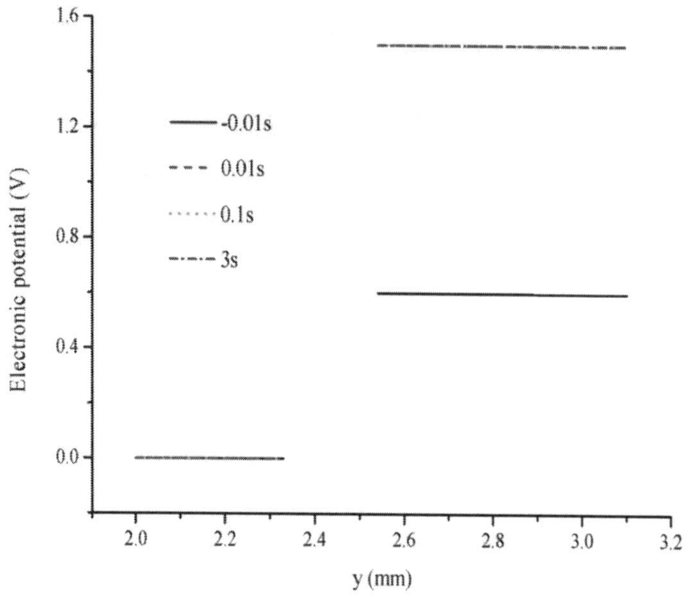

Figure 11. Evolution of electronic potential with time along line A–A in operation mode switching.

Figure 12 shows the evolution of the electrolyte potential with time along line A–A in operation mode switching. At −0.01 and −0.1 s, the cell is in the FC mode. Notably, the electrolyte potential is identical in the FC mode and is approximately −0.2 V at the interface of the CL and membrane of the oxygen electrode. Meanwhile, the electrolyte potential increases linearly from the oxygen electrode to the hydrogen electrode along line A–A and reaches the maximum at approximately 0 V at the hydrogen electrode/membrane interface. At 0 s, the overall value of the electrolyte potential increases considerably. However, the value remains negative. Once the mode is switched from the FC mode to the WE mode, the electrolyte potential changes immediately to the positive value. Then, the electrolyte potential achieves a maximum of approximately 0.15 V at the interface of the CL and membrane of the oxygen electrode by increasing linearly from the hydrogen electrode to the oxygen electrode.

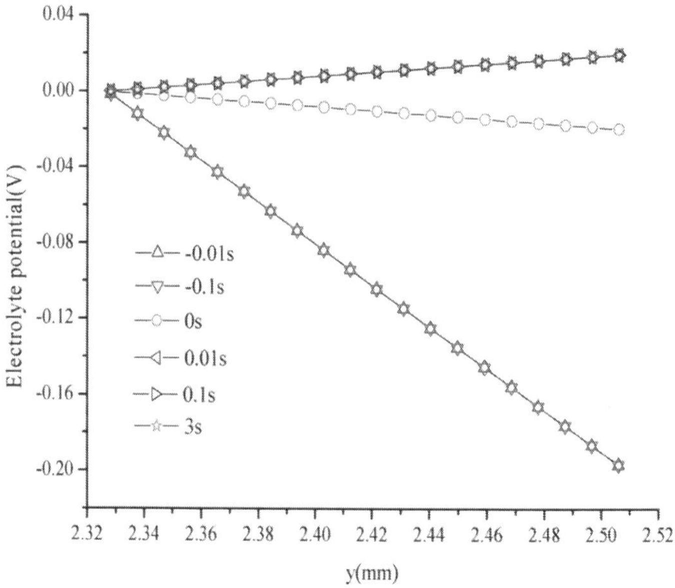

Figure 12. Evolution of electrolyte potential with time along line A–A in operation mode switching.

5. CONCLUSIONS

A 2D, single-phase, isothermal, multicomponent, transient model coupled with an electrochemical reaction is built for URFCs, which switch from the FC mode to the WE mode.

1 The distributions of parameters, such as hydrogen, oxygen, water mass fractions, and electrolyte potential, respond to the operating voltage leap via a sudden change under operation mode switching.

2 The hydrogen mass fraction gradients are smaller than the oxygen mass fraction gradients for the same scale legend.

3 Electronic potential exhibits different trends compared with other parameters. At the hydrogen electrode, the electronic potential is maintained at zero in the switching mode. At the oxygen electrode, the electronic potential is maintained at approximately 0.6 V in the FC mode and is switched to approximately 1.5 V in the WE mode. The electrolyte potential increases linearly from the oxygen/hydrogen electrode to the hydrogen/oxygen electrode in the FC/WE mode.

4 The average mass fractions of the reactants (O_2 and H_2) and product (H_2O) exhibit evident differences between each layer in the steady state of the FC mode. By contrast, the average mass fractions of the reactant (H_2O) and products (O_2 and H_2) exhibit only a slight difference between each layer in the steady state of the WE mode.

5 The duration of the switch from the transient state to the steady state in either the FC mode or the WE mode is only approximately 0.2 s.

The simulation results presented in this study will help improve our understanding of the internal transport phenomena of URFCs under operation mode switching.

ACKNOWLEDGMENTS

The authors are grateful to the National Natural Science Foundation of China (Grant No. 51476003) for the financial support.

AUTHOR CONTRIBUTIONS

Lulu Wang built the model, organized the data and wrote the main body of the paper. Hang Guo analyzed the numerical results and revised the manuscript. Fang Ye proposed the idea of modeling. Chongfang Ma supervised the research process. All authors read and approved the manuscript.

CONFLICTS OF INTEREST

The authors declare no conflict of interests.

REFERENCES

1. Verma, A.; Basu, S. Feasibility study of a simple unitized regenerative fuel cell. *J. Power Sources* **2004**,*135*, 62–65.
2. Millet, P.; Ngameni, R.; Grigoriev, S.A. Scientific and engineering issues related to PEM technology: Water electrolysers, fuel cells and unitized regenerative systems. *Int. J. Hydrog. Energy* **2011**, *36*, 4156–4163.
3. Mitlitsky, F.; Myers, B.; Weisberg, A.H. Reversible (unitized) PEM fuel cell devices. *Fuel Cells Bull.* **1999**, *11*, 6–11.
4. Grigoriev, S.A.; Millet, P.; Porembsky, V.I. Development and preliminary testing of a unitized regenerative fuel cell based on PEM technology. *Int. J. Hydrog. Energy* **2011**, *36*, 4164–4168.
5. Applyby, A.P. Regenerative fuel cells for space applications. *J. Power Sources* **1988**, *22*, 377–385.
6. Markgraf, S.; Horenz, M.; Schmiel, T. Alkaline fuel cells running at elevated temperature for regenerative fuel cell system applications in spacecrafts. *J. Power Sources* **2012**, *201*, 236–242.

7. Yoshitsugu, S. A 100-W class regenerative fuel cell system for lunar and planetary missions. *J. Power Sources* **2011**, *196*, 9076–9080.

8. Guarnieri, M.; Alotto, P.; Moro, F. Modeling the performance of hydrogen-oxygen unitized regenerative proton exchange membrane fuel cells for energy storage. *J. Power Sources* **2015**, *297*, 23–32.

9. Herrera, O.E.; Wilkinson, D.P.; Merida, W. Anode and cathode overpotentials and temperature profiles in a PEMFC. *J. Power Sources* **2012**, *198*, 132–142.

10. Zhan, Z.G.; Wang, C.; Fu, W.G. Visualization of water transport in a transparent PEMFC. *Int. J. Hydrog. Energy* **2012**, *37*, 1094–1105.

11. Nguyen, T.V.; White, R.E. A water and heat management model for proton-exchange-membrane fuel cells. *J. Electrochem. Soc.* **1993**, *140*, 2178–2186.

12. Ramousse, J.; Deseure, J.; Lottin, O. Modeling of heat, mass and charge transfer in a PEMFC single cell. *J. Power Sources* **2005**, *145*, 416–427.

13. Hu, G.L.; Fan, J.R. A three-dimensional, multicomponent, two-phase model for a proton exchange membrane fuel cell with straight channels. *Energy Fuels* **2006**, *20*, 738–747.

14. Singh, D.; Lu, D.M.; Djilali, N. A two-dimensional ananlysis of mass transport in proton exchange membrane fuel cells. *Int. J. Eng. Sci.* **1999**, *37*, 431–452.

15. Marangio, F.; Santarelli, M.; Cala, M. Theoretical model and experimental analysis of a high pressure PEM water electrolyser for hydrogen production. *Int. J. Hydrog. Energy* **2009**, *34*, 1143–1158.

16. Nie, J.H.; Chen, Y.T. Numerical modeling of three-dimensional two-phase gas-liquid flow in the flow field plate of a PEM electrolysis. *Int. J. Hydrog. Energy* **2010**, *35*, 3183–3197.

17. Carmo, M.; Fritz, D.L.; Mergel, J. A comprehensive review on PEM water electrolysis. *Int. J. Hydrog. Energy* **2013**, *38*, 4901–4934.

18. Grigoriev, S.A.; Kalinnikov, A.A.; Millet, P. Mathematical modeling of high-pressure PEM water electrolysis. *J. Appl. Electrochem.* **2010**, *40*, 921–932.

19. Jung, H.Y.; Huang, S.Y.; Popov, B.N. High-durability titanium bipolar plate modified by electrochemical deposition of platinum for unitized regenerative fuel cell(URFC). *J. Power Sources* **2010**, *195*, 1950–1956.

20. Chen, G.B.; Zhang, H.M.; Zhong, H.X. Gas diffusion layer with titanium carbide for a unitized regenerative fuel cell. *Electrochim. Acta* **2010**, *55*, 8801–8807.

21. Pai, Y.H.; Tseng, C.W. Preparation and characterization of bifunctional graphitized carbon-supported Pt composite electrode for unitized regenerative fuel cell. *J. Power Sources* **2012**, *202*, 28–34.

22. Huang, S.Y.; Ganesan, P.; Jung, H.Y. Development of supported bifunctional oxygen electrocatalysts and corrosion-resistant gas diffusion layer for unitized regenerative fuel cell applications. *J. Power Sources* **2012**, *198*, 23–29.

23. Lee, W.H.; Kim, H. Optimization of electrode structure to suppress electrochemical carbon corrosion of gas diffusion layer for unitized regenerative fuel cell. *J. Electrochem. Soc.* **2014**, *161*, 729–733.

24. Gabbasa, M.; Sopian, K.; Fudholi, A. A review of unitized regenerative fuel cell stack: Material, design and research achievements. *Int. J. Hydrog. Energy* **2014**, *39*, 17765–17778.

25. Doddathimmaiah, A.; Andrews, J. Theory, modeling and performance measurement of unitized regenerative fuel cells. *Int. J. Hydrog. Energy* **2009**, *34*, 8157–8170.

26. Hoberecht, M.A.; Robert, D.G. Use of excess solar array power by regenerative fuel cell energy storage systems in low earth orbit. In Proceedings of the IEEE Energy Conversion Engineering Conference, Honolulu, HI, USA, 27 July–1 August 1997.

27. Jin, X.F.; Xue, X.J. Mathematical modeling analysis of regenerative solid oxide fuel cells in switching mode conditions. *J. Power Sources* **2010**, *195*, 6652–6658.

28. Raj, A.; Shamim, T. Investigation of the effect of multidimensionality in PEM fuel cells. *Energy Convers. Manag.* **2014**, *86*, 443–452.

29. Ju, H.; Wang, C.Y. Experimental Validation of a PEM fuel cell model by current distribution data. *J. Electrochem. Soc.* **2004**, *151*, A1954–A1960.

30. Dihrab, S.S.; Razali, A.M. Studies on a single cell unitized regenerative fuel cells. In Proceedings of the 9th ESEAS International Conference on System Science and Simulation In Engineering, Iwate, Japan, 4–6 October 2010.

31. Rabih, S.; Rallieres, O.; Turpin, C.; Astier, S. Experimental Study of a PEM Reversible Fuel Cell. 2011. (accessed on 17 July 2015).

32. Doubek, G.; Robalinho, E.; Cunha, E.F. Application of CFD techniques in the modeling and simulation of PBI PEMFC. *Fuel Cells* **2011**, *11*, 764–774.

33. Sehribani, U. Mathematical and Computational Modeling of Polymer Exchange Membrane Fuel Cells. Master's Thesis, University of Nevada, Reno, NV, USA, August 2012.

34. Hsuen, H.K.; Yin, K.M. Performance equations of proton exchange membrane fuel cells with feeds of varying degree of humidification. *Electrochim. Acta* **2012**, *62*, 447–460.

35. Ni, M. Computational fluid dynamics modeling of a solid oxide electrolyzer cell for hydrogen production. *Int. J. Hydrog. Energy* **2009**, *34*, 7795–7806.

36. Khazaee, I. Experimental investigation and numerical comparison of the performance of a proton exchange membrane fuel cell at different channel geometry. *Heat Mass Transf.* **2015**, *51*, 1177–1187.

CHAPTER 5

Enhanced Mass Transfer in Microbubble Driven Airlift Bioreactor for Microalgal Culture

Kezhen Ying[1*], Mahmood K. H. Al-Mashhadani[1], James O. Hanotu[1], Daniel J. Gilmour[2], William B. Zimmerman[1]

[1]Department of Chemical and Biological Engineering, University of Sheffield, Sheffield, UK

[2]Department of Molecular Biology and Biotechnology, University of Sheffield, Sheffield, UK

ABSTRACT

In this study, the effect of microfluidic microbubbles on overall gas-liquid mass transfer (CO_2 dissolution and O_2 removal) was investigated under five different flow rates. The effect of different liquid substrate on CO_2 mass transfer properties was also tested. The results showed that the K_La can be enhanced by either increasing the dosing flowrate or reducing the bubble size; however, increasing the flow rate to achieve a higher K_La would ultimately lower the CO_2 capture efficiency. In order to achieve both higher CO_2mass transfer rate and capture efficiency, reducing bubble size (e.g. using microbubbles) has been proved more promising than increasing flow rate. Microbubble dosing with 5% CO_2 gas showed improved K_La by 30% - 100% across different flow rates, compared to fine-bubble dosing. In the real algal culture medium, there appears to be two distinct stages in terms of K_La, divided by the pH of 8.4.

KEYWORDS

Microbubbles; K_La; CO_2 Capture; Algal Culture

1. INTRODUCTION

The cultivation of microalgae has been studied and developed for more than 40 years [1]. Two of the major limiting factors for microalgal culture are light and CO_2 as they are the key participants for the "light reactions" and "dark reactions" in photosynthesis, respectively. Many researches have been carried out to study the impact of light on algal growth. Technical issues associated with light have been also well studied especially for photobioreactos, with various solutions (e.g. using an optimal mixing rate and light/dark ratio, combining artificial light with natural light, and increasing harvest frequency etc.) [2]. As regards to CO_2 supply, in most microalgae cultures, CO_2 is usually injected into the culture through bubbling CO_2 enriched air into porous diffusers, which promises a gas transfer efficiency of 13% - 20% [3]. Additional supply of CO_2 contributes many benefits to the culture. First of all, the supply of CO_2 can lead to enhanced algal metabolisms, and on the other hand, it can act as buffer solution to neutralize the increased pH caused by algal growth. Secondly, supply of CO_2 enhances the internal mixing of bioreactor, helping to evenly distribute nutrients and the exposure time of algal cells to light. Furthermore, accumulation of O_2 in culture medium is toxic to microalgal cells and it is one of the major limiting factors for scale up of the bioreactor [4]. Introducing CO_2 into culture also helps to strip accumulated oxygen and hence prevents algal cells from toxicity [5]. According to the relationship between partial pressure and Gibbs free energy (Equation (1)), it is found that the increase in the partial pressure of reactants (e.g. CO_2) or the reduction of partial pressure in the products (e.g. O_2) results in the value of Gibbs free energy becoming negative. Hence the reaction becomes thermodynamically favourable and moves towards to the formation of more products [6]. Such feature of performance is widely utilized for many bioprocesses to achieve a higher productivity [2,5]. Therefore by increasing the concentration of dissolved CO_2 whilist reducing the accumulated O_2 level can be considered as an approach towards improving productivity.

$$aA + bB \rightarrow cC + dD$$

$$\Delta G = \Delta G^{\circ} + RT \ln \frac{[p_C]^c [p_D]^d}{[p_A]^a [p_B]^b} \qquad (1)$$

However, most existing CO_2 supply techniques are relatively inefficient. Due to low interfacial surface area between gas bubbles and culture medium, the gas-liquid mass transfer is poor, which associated with CO_2 loss to atmosphere [7]. Besides, additional CO_2 supply increases the operational cost, which can not be balanced eventually by the algal yields enhancement due to the low CO_2 mass transfer. Improving the CO_2 supply efficiency and consequently enhancing the algal productivity has become a major challenge over the years. Design of bioreactor with low energy cost and particularly high gas mass transfer for both CO_2 dissolution and O_2 removal is hence a major consideration for cost-competitive microalgae culture.

Due to the enhanced gas-liquid mass transfer efficiency and liquid circulation etc., airlift bioreactors (ALB) are increasingly employed for microalgae culture. Many researches have been carried out on the performance of different ALBs; however, these studies were carried out all based on conventional gas supply system. There are few studies on the effects of microbubbles on ALB performance, because normally the microbubble generation systems, for instance DAF, electro-flotation, electrostatic spraying, and mechanical agitation etc, were not profitable to be applied into most bio-processes due to their high energy cost [8-12]. Recently, an innovative microbubble generation system (fluidic oscillator) with lower power consumption has been invented by Zimmerman et al. [13,14] with the benefits of energy saving and improved efficiency. The detailed information on fluidic oscillator and its microbubble generation mechanism were described in previous studies [13-15]. This study aims to investigate the effect of microbubbbles (generated by fluidic oscillator) on mass transfer under different aeration flow rates. In addition, the impact of different liquid substrate (e.g. $NaHCO_3$ medium and algae medium) on CO_2 mass transfer properties is investigated.

2. MATERIALS AND METHODS

2.1. Materials

A 7L-airlift loop bioreactor based on classic ALB geometry designs [16], as shown in **Figure 1** (left), was used to study the mass transfer properties of microbubbles and fine-bubbles. Besides, a smaller version (3L) of ALB, based on the similar geometry design, was applied to study the impact of different liquids on mass transfer, shown in **Figure 1** (right).

2.2. Experimental Procedure

To study the mass transfer properties of microbubbles and fine-bubbles, the two inlet ports of diffusers at the bottom of bioreactor were connected to gas cylinder by PVC tubes, through a fluidic oscillator or a Y-junction.

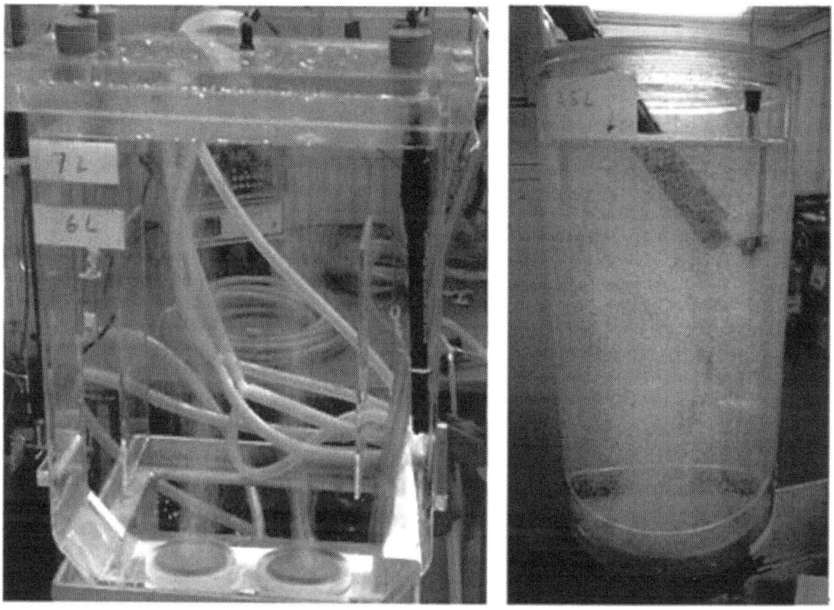

Figure 1. 7L (left) and 3L (right)—lab scale airlift loop bioreactors (ALB).

The 7 L-airlift bioreactor is made of transparent acrylic material with the dimension of 26 cm × 10 cm × 30 cm. Inside the bioreactor, two ceramic diffusers (d = 5 cm, h = 1 cm), with the pore size of 20 microns, are fixed at the bottom. Two draught baffles are suspended 3.5 cm above these diffusers, dividing the middle chamber into 3 regions which work as risers and downcomers. A static liquid height of 15 cm was employed to give the volume of 7 L. There are several holes drilled on the lid to allow pH and DO probes insert into the reactor. The 3Lairlift bioreactor is also made of acrylic material, with the dimension of 285 mm in height and 124 mm in diameter. The air lift loop design consists of a ceramic diffuser (diameter of 78 mm and pore size of 20 microns) fixed at bottom and an internal draught tube (H: 170 mm, D: 95 mm) hanged 30 mm above the diffuser.

The detailed connections for main experiments (with oscillator) and control experiments (without oscillator) are illustrated in **Figure 2**. A pH and DO probe (Mettler Toledo, UK) were inserted into the bioreactor via the holes on the lid. These holes were blocked by rubber bungs to prevent gas leakage. The outlet nozzle on the lid was connected to a flow meter to measure the outlet flowrate which is equal to the real inlet flowrate. For each set of experiment, 7L distilled water with the temperature 25˚C ± 1˚C were employed. Mixture gas containing 5% CO_2 and 95% N_2 was injected into bioreactor under certain flow rate. Five different flow rates were tested. The flow rate was measured by a flow meter which was connected to the outlet port of the bioreactor. The changes in pH and DO were monitored by pH meter and DO meter respectively. Data was recorded every 30 seconds until pH and DO readings were stable. The effect of different liquids on mass transfer was studied in the 3L-ALB with the same setup as shown in **Figure 2**. The mass transfer for CO_2dissolution was tested in the distilled water containing certain concentration of $NaHCO_3$ and also in the real algal culture medium (containing algae). 7 different concentrations of $NaHCO_3$ were tested. The algae (Dunaliella salina) used in this study was 7 days old. During the mass transfer test, 5% CO_2 and 95% N_2 was injected into D. Salina culture under a fixed dosing flow rate (0.7L/min), with DO and pH recorded every 30 seconds. The dissolved CO_2 concentration was calculated based on Equation (2) (for water)

For microbubble dosing, the gas ejected from cylinder flowed into a fluidic oscillator, and was shot out from the two outlet terminals on oscillator. Here a flow rate of 80 L was required to drive the oscillator. Before such amount of gas injected into bioreactor, most of them were bled out via bleeding pipes with only less than 1% flowing into bioreactor, and the real inlet flow was measured by the flow meter at bioreactor output. For finebubble dosing, the area marked as red frame was placed with a Y-junction or Equation (3) (for $NaHCO_3$ medium and algal medium) [17]. $[Na^+]$ in Equation (3) particularly means the concentration of Na^+ obtained from $NaHCO_3$. The method of mass transfer coefficient estimation was estimated as the slope of a semilog plot of 1/(1-E) versus T, which was explained in details in Chisti (1989) [16].

$$[CO_2] = \frac{\left(10^{-pH} - 10^{pH-14}\right)10^{-2pH}}{10^{-6.381-pH} + 2\times 10^{-16.758}}(mol/L) \quad (2)$$

$$[CO_2] = \frac{\left(10^{-pH} - 10^{pH-14} + \Delta\left[Na^+\right]\right)10^{-2pH}}{10^{-6.381-pH} + 2\times 10^{-16.758}}(mol/L) \quad (3)$$

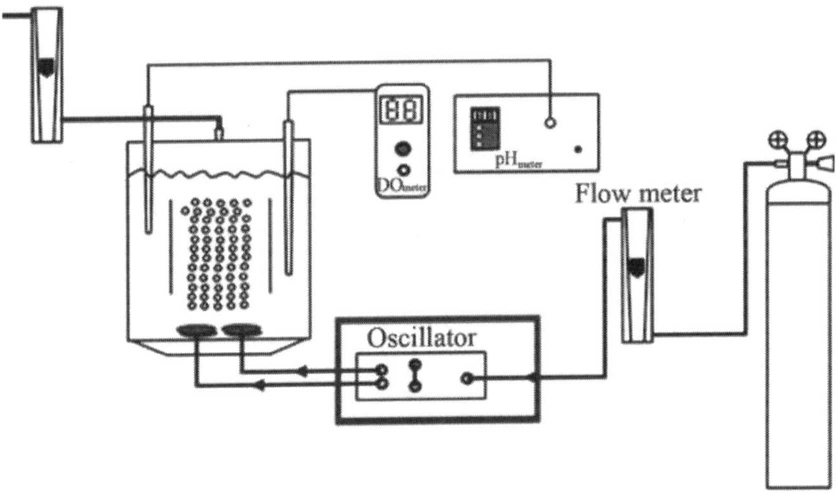

Figure 2. Connections for mass transfer test.

3. RESULTS AND DISCUSSIONS

3.1. Mass Transfer for Microbubble Driven and Fine-Bubble Driven Reactor

The effects of microbubble dosing on mass transfer for CO_2 dissolution and O_2 removal were examined by dosing 5% CO_2 mix-gas (balanced with 95% N_2) into bioreactor (containing 7L distilled water) under 5 different bubbling flow rates, along with the control experiment (without fluidic oscillator, fine bubble dosing). The mass transfer coefficient K_La for CO_2 dissolution and O_2removal under each bubbling condition were plotted in **Figure 3**. From **Figure 3**, generally K_La for either CO_2 dissolution or O_2 stripping increases along with gas dosing flow rate. For K_La, K_L mainly depends on the gasliquid properties (e.g. density, viscosity, diffusivity and temperature etc.), and therefore is usually considered as a constant for the fixed circumstance [16]. Chisti expressed the interfacial area "a" as a function of gas holdup (ε) and bubble diameter (d_B) [16], shown as:

$$a = \frac{6\varepsilon}{d_B} \tag{4}$$

FO means "with fluidic oscillator", representing microbubble (300 - 400 µm) dosing while NoFO stands for "without fluidic oscillator", representing fine bubble (500 - 600 µm) dosing. Due to the lab limitations, the error bars shown in this figure was obtained from the duplication of fine bubble dosing under each flow rate (for O_2 removal, NoFO).

For the same bubbling system (either microbubble dosing or fine bubble dosing), the bubble size can be considered as the same for different gas dosing flowrates, while the gas holdup usually increases with the bubbling flowrate, therefore, K_La was enhanced by increasing the flowrate as the total interfacial area was amplified.

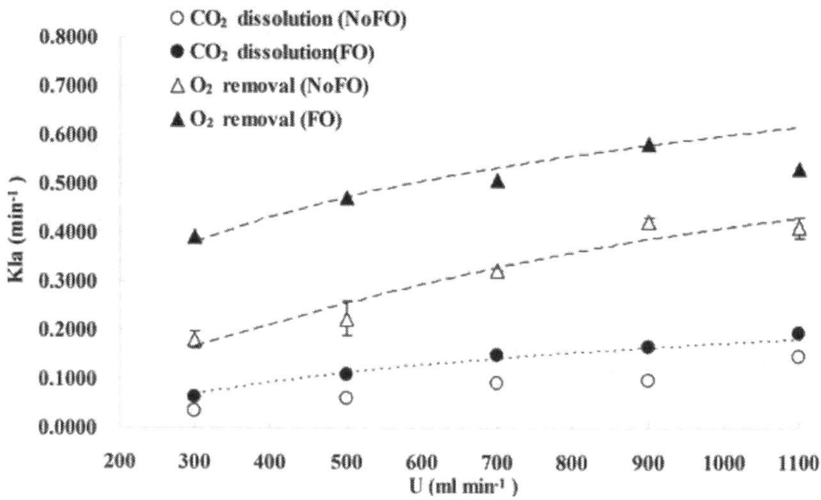

Figure 3. Mass transfer coefficients under different dosing conditions.

As regard the comparison of mass transfer coefficients (either for CO_2 dissolution or O_2 removal) between microbubble dosing (FO) and fine bubble dosing (NoFO), microbubbles had a higher mass transfer coefficient under each dosing flowrate. For CO_2 dissolution, the highest K_La under fine bubble dosing (0.14 min^{-1}) was achieved at dosing flow rate of 1.1 L×min^{-1}, while almost the same K_La value (0.15 min^{-1}) was achieved by microbubble dosing at 0.7 L·min^{-1}. Similarly, for O_2removal the highest K_La (0.41 min^{-1}) was obtained at 1.1 L·min^{-1} by fine bubble dosing, which however can be achieved at only 0.3 L×min^{-1}. The potential for energy saving, especially for large scale processes, is therefore straight forward to argue. For example, in order to dissolve more $CO_2$and strip off O_2 accumulated in algal bioreactor, it would typically require dosing certain percentage of CO_2 mixture gas at a relatively high aeration rate to achieve a desired mass transfer coefficient. But actually, under very high flow rate, most of the gas is wasted. And an intensive agitation caused by high flow rate may damage the algal cells. However, by using oscillator (microbubble dosing) the desired K_La can be obtained even at relatively low flow rate. It considerably saves gas usage and also the electricity cost.

In conclusion, for the same bubble generation method (the changes in bubble sizes are considered to be negligible across a wide range of dosing flow rate), enhancing the gas dosing flowrate (which means enhancing the gas hold up for the same liquid volume) can increase the mass transfer coefficient. For the same bubbling flowrate, reducing the bubble size can lead to an improvement on K_La as well. In another word, K_La can be enhanced by either increasing the dosing flowrate (to be more accurate, flowrate/liquid volume ratio) or reducing the bubble size.

3.2. The Improvement of K_La by Using Fluidic Oscillator

When injecting CO_2/N_2 mixture gas into water, CO_2 dissolution happens along with O_2 stripping. The improvements by using fluidic oscillator (microbubbles) on mass transfer for CO_2 dissolution and O_2 stripping can be simply quantified as the percentage increase in $K_{L(CO_2)}$a and $K_{L(O_2)}$a, expressed in Equation (5) and Equation (6), respectively.

$$I_{K_{L(CO_2)}a}\% = \frac{K_{L(CO_2)}a_{FO} - K_{L(CO_2)}a_{NoFO}}{K_{L(CO_2)}a_{NoFO}} \tag{5}$$

$$I_{K_{L(O_2)}a}\% = \frac{K_{L(O_2)}a_{FO} - K_{L(O_2)}a_{NoFO}}{K_{L(O_2)}a_{NoFO}} \tag{6}$$

Either Equation (5) or Equation (6) can be turned into Equation (7) which indicates the percentage increase in K_La by using microbubble dosing should be the same for either CO_2 dissolution or O_2removal under a fixed bubbling flowrate. The percentage improvement of K_La is therefore determined by the percentage difference of the total interfacial areas for a certain dosing flow rate. Combining Equation (7) and Equation (4), the percentage increase in K_La is correlated to bubble diameter (d_B) and gas hold-up (ε), described by Equation (8).

$$I\% = I_{K_{L(CO_2)}a}\% = I_{K_{L(O_2)}a}\% = \frac{a_{FO} - a_{NoFO}}{a_{NoFO}} \tag{7}$$

$$I\% = \frac{\varepsilon_{FO} d_{BNoFO}}{\varepsilon_{NoFO} d_{BFO}} - 1 \tag{8}$$

From Equation (8), the efficiency of K_La improvement (I%) therefore should be the same across different flow rates, assuming 1) the gas holdups are identical between microbubble dosing and fine bubble under the same flow rates, and 2) changing the flow rate dose not vary the average bubble size for either microbubbles or fine bubbles as long as the "bubble coalescence" does not happen. However, the experimental results are inconsistent with such speculation. **Figure 4** shows the K_La percentage increase. In general, microbubble dosing enhances the K_La by 30% - 100% over a wide flow rate range, while the efficiency of the improvement decreases when increasing the flow rate. It is speculated that the microbubble size increases with the flow rate. The fluidic oscillator provides a periodical oscillating pulse to neck-off the bubbles attached to the diffuser orifice when they are still small. But for the same surface area of diffuser, increasing the flow rate may change the oscillating properties (e.g. the attenuation of "pulse force" due to the build up of boundary layer), and may also cause bubble coalescence, consequently weakening the efficiency of oscillator for mirobubble creation. Therefore, the microbubble size may slightly increase when the flow rate increases, resulting in a reduction of d_{BNoFO}/d_{BFO} ratio, which leads to the decline of K_La improvement efficiency (I%). This phenomenon also indicates a view that using fluidic oscillator to enhance mass transfer has its limitations in terms of flow rate (or to be more accurate, flow rate over liquid volume ratio).

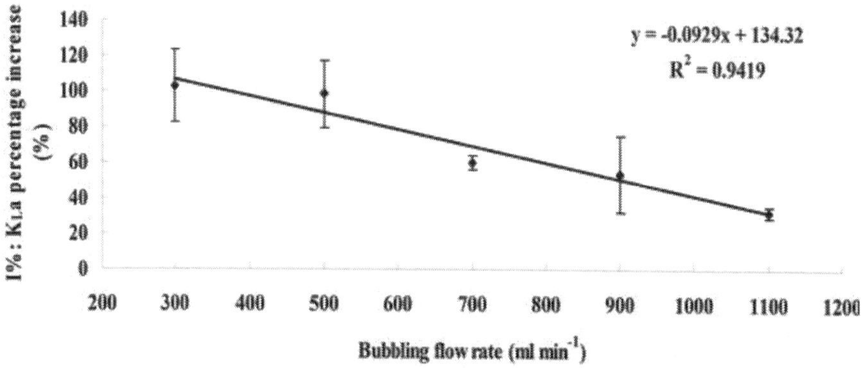

Figure 4. Plots of K_La percentage increase versus dosing flow rate.

The value of I% under each flow rate was the average between the values calculated based on Equation (5) and Equation (6). The error bar represents the standard deviation between these two values.

3.3. The Relationship between Mass Transfer Coefficient and Overall Mass Transfer Rate

Knowing the mass transfer coefficient K_La helps to indicate the capability of mass transfer, while knowing the mass transfer rate gives a straight view of e.g. how rapidly the CO_2 dissolve into liquid, which also helps to estimate the CO_2 capture efficiency.

The instantaneous mass transfer rate (V_{MTR}) is interpreted as the driving force multiplied by the mass transfer coefficient (K_La) [16], shown in Equation (9),

$$V_{MTR} = \frac{\mathrm{d}[\mathrm{CO_2}]}{\mathrm{d}t} = K_L a\left([\mathrm{CO_2}]^* - [\mathrm{CO_2}]_t\right), \quad (9)$$

where K_La is the mass transfer coefficient (min^{-1}), both $[CO_2]_t$ and $[CO_2]^*$ are instantaneous concentrations of CO_2 and its equilibrium concentration ($mol\times L^{-1}$), respecttively. The average mass transfer rate (V'_{MTR}) for a certain dosing time period (t_d) can be fairly represented as

$$V'_{MTR} = \frac{\int_0^{t_d} V_{MTR}\,dt}{t_d} = \frac{\int_0^{t_d} K_L a\left([CO_2]^* - [CO_2]_t\right)dt}{t_d} \quad . \qquad (10)$$

Assuming

$$[CO_2]_t = [CO_2]_0 + V'_{MTR}t \quad , \qquad (11)$$

by solving Equation (10) and Equation (11), it gives:

$$V'_{MTR} = \frac{\int_0^{t_d} K_L a\left([CO_2]^* - [CO_2]_0 - V'_{MTR}t\right)dt}{t_d}$$

$$= K_L a[CO_2]^* - K_L a[CO_2]_0 - K_L a V'_{MTR}\frac{t_d}{2}$$

$$\Rightarrow V'_{MTR} = \frac{K_L a\left([CO_2]^* - [CO_2]_0\right)}{\frac{K_L a t_d}{2} + 1} \qquad (12)$$

where $[CO_2]_0$ represents the initial CO_2 concentration ($mol\times L^{-1}$) for a selected time period.

The accuracy of Equation (12) was examined via Figure 5 which plots the experimental values of average mass transfer rates versus the calculated values by using Equation (12). Compared with examined values, most of the data calculated by Equation (12) showed only less than 10% difference.

3.4. CO_2 Capture Efficiency for Microbubble Dosing and Fine-Bubble Dosing

CO_2 capture efficiency is one of the most important parameter concerned by many bioprocesses with the purpose of CO_2 sequestration. Since the rate of CO_2 dissolving into liquid can be valuated by overall mass transfer rate using Equation (12), the CO_2 capture efficiency $\left(E_{CO_2}\right)$ can be therefore simply described as the amount of CO_2 been absorbed over the amount of CO_2 been fed into the liquid (m_s/m_d) within a specific dosing time period (t_d), shown in Equation (13).

$$E_{CO_2} = \frac{m_s}{m_d} = \frac{V'_{MTR} \times Vol \times t_d}{CO_2\% \times V_F \times P/(RT) \times t_d} \tag{13}$$

where $CO_2\%$ means the percentage of CO_2 in the gas supply, Vol is the volume of the liquid (m^3), V_F is the gas dosing flow rate ($L \times min^{-1}$), P is standard atmosphere pressure (101,325 Pa), R is the ideal gas law constant (8.314 $J \times K^{-1} \times mol^{-1}$) and T is the temperature (298 K).

The CO_2 dissolving rate and the CO_2 capture efficiency under different dosing conditions were plotted in Figures 6 and 7, respectively. In general, microbubble dosing by using the fluidic oscillator was found to have both higher CO_2 dissolving rate (average mass transfer rate) and CO_2sequestration efficiency for a wide range of dosing flow rate, but the level of improvements were attenuated as the flow rate went up (similar to the attenuation of K_La improvement, see 3.2). Such attenuation of improvement was caused by the increase in microbubble size due to the weakening of oscillation and bubble coalescence under higher flow rate.

Apart from reducing bubble size, increasing flow rate can also achieve a higher K_La (see 3.1), it is therefore not a surprise to found that the CO_2 overall mass transfer rate increases along with the flow rate (**Figure 6**). However,

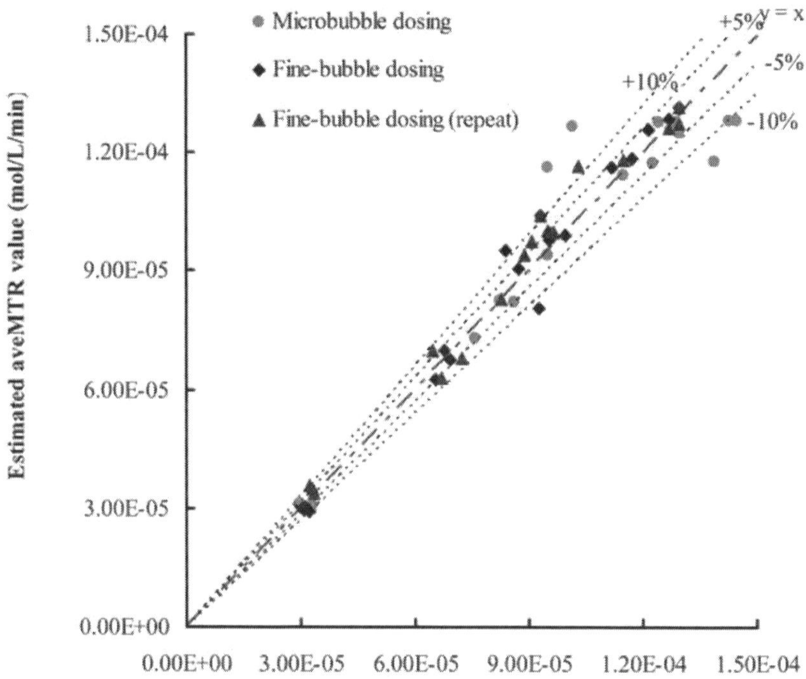

Figure 5. Plots of estimated average mass transfer rates versus examined values.

Experimental aveMTR value(mol/L/min)

The K_La value for each condition was obtained based on selected time period (5 - 10 min), via the standard method described by Chisti [16]. For each dosing condition, the average mass transfer rates for selected time period (5 - 8 min, 5 - 10 min and 5 - 12 min) were estimated by Equation (12) (Y-axis) and examined by $([CO_2]_t - [CO_2]_0)/t$ (X-axis).

it is interesting that the CO_2 capture efficiency actually reduces when the flow rate increases (**Figure 7**). Higher K_La dose mean higher CO_2 overall mass transfer rate (higher CO_2 dissolving rate), however, if the cost to achieve higher K_La is enhancing the dosing flow rate rather than reducing bubble size, then the amount of not dissolved CO_2 ("wasted CO_2") would increase, and such increase in wasted CO_2 could not be balanced by the increase in dissolved CO_2, which ultimately lowers the CO_2 capture efficiency. Therefore, in order to achieve both higher CO_2 mass transfer rate and capture efficiency, reducing bubble size (e.g. using microbubbles) is more promising than increasing flow rate.

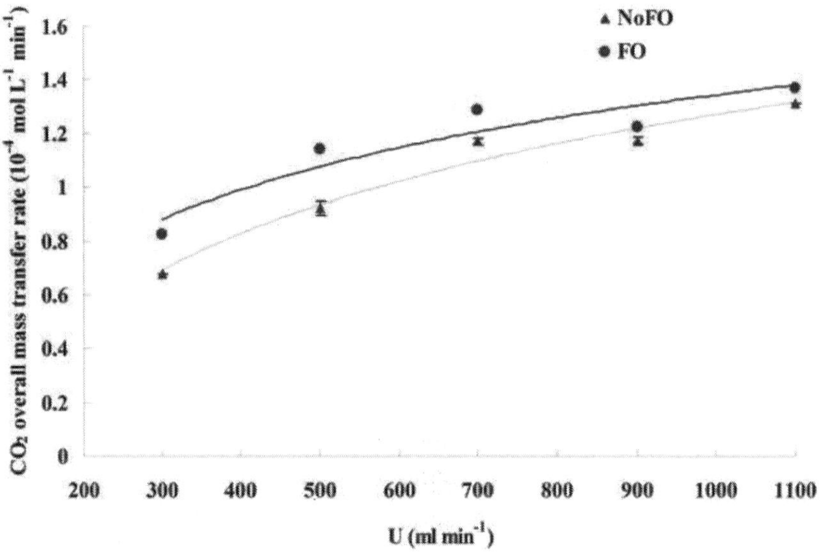

Figure 6. The average mass transfer rate under different dosing conditions.

The average mass transfer rate was calculated based on Equation (12), and the time period selected for MTR' calculation under each dosing condition was between 5 min and 10 min after starting dosing, the same time interval used for the K_La estimation.

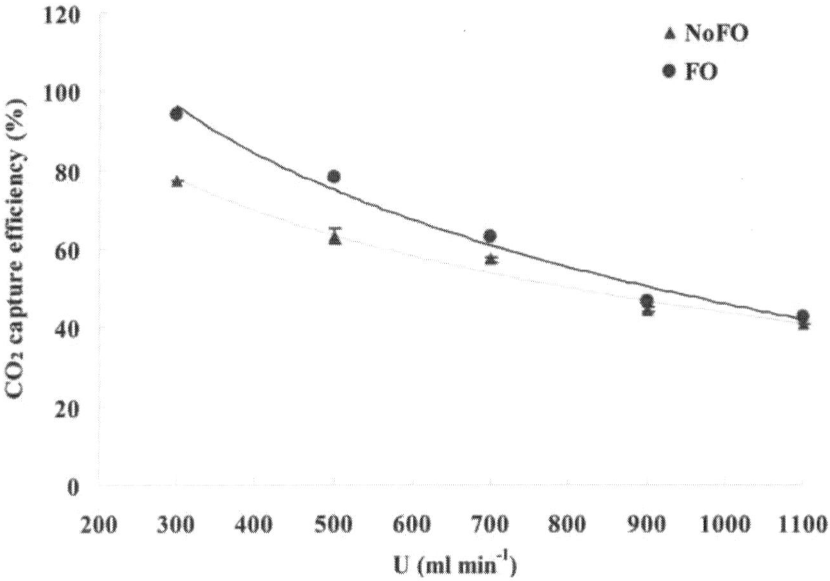

Figure 7. The plots of CO_2 capture efficiency versus gas dosing flowrate.

The CO_2 capture efficiency for each dosing condition was calculated based on Equation (13), and the time period selected for each calculation under different dosing conditions is the same as for overall mass transfer rate calculation.

3.5. Effect of $NaHCO_3$ on Equilibrium pH and CO_2 Mass Transfer Rate in Water

In microalgae culture, CO_2 is injected into the culture medium (usually containing $NaHCO_3$) rather than pure water. When adding $NaHCO_3$ into water, $NaHCO_3$ dissociates into sodium (Na^+) and bicarbonate (HCO_3^-) ions, and these HCO_3^- ions neutralize some of the H^+ ions present in the medium to form the dissolved CO_2 and so increase the pH. So the concentration of $NaHCO_3$ clearly has an effect on pH, it is worth finding out whether the culture medium containing $NaHCO_3$ could affect the

CO_2 mass transfer. Therefore, a separate experiment was carried out in a smaller version but the same design of airlift bioreactor (2.5 L).

Keeping other parameter constant (flow rate, temperature etc.), higher concentrations of $NaHCO_3$added into distilled water should theoretically raise the minimal pH (equilibrium pH, pH^*) reached after CO_2 dosing. According to Henry's law and Two-film theory, the equilibrium concentration of dissolved CO_2 ($[CO_2]^*$) should only depend on the CO_2 partial pressure in the gas phase for a fixed gas/liquid properties and temperature (assuming the changes in liquid physical properties by adding different amount of $NaHCO_3$ to the water, e.g. viscosity, are negligible, as long as the concentration of $NaHCO_3$ is low). Therefore, different concentrations of $NaHCO_3$ in the water should not affect the $[CO_2]^*$. On the other hand, the CO_2 concentration is correlated to pH by Equation (3). Since the concentration of Na^+ varies for different concentration of $NaHCO_3$, while the $[CO_2]^*$ does not change, it is therefore reasonable that pH^* changes for the water containing various $NaHCO_3$concentration. Indeed, this hypothesis was proved correct, in **Figure 8**, a log-linear trend was observed in the equilibrium pH values as the concentration of $NaHCO_3$ was increased. Besides, all the equilibrium concentrations of CO_2 corresponding to each equilibrium pH value under different concentrations of $NaHCO_3$ were found to be the same, which is approximately 0.0017 ± 0.0001 mol·L^{-1}. In terms of mass transfer for CO_2 dissolution, it can be seen that changing the concentration of $NaHCO_3$ does not have much of an effect on the mass transfer coefficient (**Figure 9**). Hence, $NaHCO_3$ could be used to control the equilibrium (minimal) pH of the medium without affecting the $[CO_2]^*$ and K_La. The pH region can also be altered depending on the particular strain of microalgae being cultured, as different algae prefer different pH.

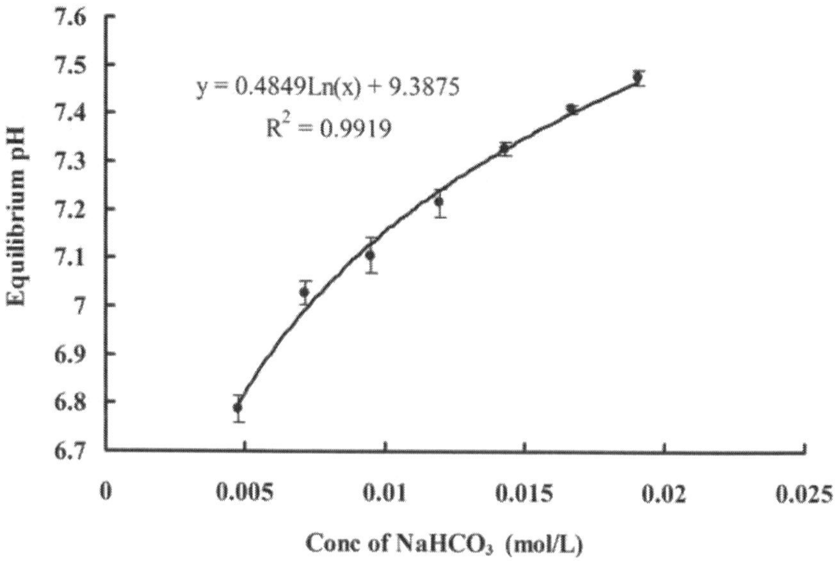

Figure 8. Change in equilibrium (final) pH for different concentration of $NaHCO_3$.

3.6. CO_2 Mass Transfer in Microalgae Culture

In order to test the effect of real microalgae culture on CO_2 mass transfer, 5% CO_2 was dosed into a healthy D. Salina culture (containing 0.012 mol/L of $NaHCO_3$) under a fixed flow rate (0.7 L/min) for 30 min, with pH recorded every 30 seconds. The results showed that there appears to be two distinct stages in terms of K_La, see **Figure 10** for example. The calculations leading to the determination of the K_La mass transfer coefficients from the slopes seen in **Figure 10** are given inTable A1 (See Appendix).

From the calculations in Table A1, for the first 4.5 minutes of gas supply, the concentration of CO_2dissolved and the resultant mass transfer is of different magnitudes when the pH is greater than 8.4. Once the pH is less than 8.4, the K_La is much higher in comparison. This was observed for each mass transfer test in culture medium with the threshold pH value of 8.4 seen each time. It is speculated that slower mass transfer at the start when pH is higher than 8.4 could be due to the chemical reactions taking place within the culture medium the gas is being bubbled through. Considering the dissociation of water into hydrogen (H^+) and hydroxyl (OH^-) ions, when the pH is over 8.4, the concentration of hydroxyl ions will be much greater than that of the hydrogen ions $([OH^-] \ll [H^+])$. The $[H^+]$ produced when CO_2 dissolves will be neutralized by $[OH^-]$ present in the medium. Considering the carbonate equilibrium system (Equation (14)) [18], this will result in less dissolved CO_2 and instead, more HCO_3^-. Hence, it is argued that the mass transfer for dissolved CO_2 will be low at relatively high pH as most of the supplied CO_2 will react to form the bicarbonate species rather than the desired dissolved CO_2. At less alkaline pH values, this transfer will be less significant and dissolved CO_2 concentration will increase faster. This theory is also applicable to the mass transfer within a medium without microalgae, but such alkaline pH values were not encountered in the experiments conducted. Stemler (1980) also noted the effect of pH on the amount of dissolved CO_2 and discusses the effect of pH on the relative levels of CO_2 and HCO_3^- present within a solution [19]. He found that in going from pH 8.0 to 7.3, the amount of HCO_3^- changed very little while the concentration of CO_2, on the other hand, increased more than 4-fold.

$$\begin{aligned} &CO_{2(gas)} + H_2O \leftrightarrow H_2CO_{3(aq)} \\ &\leftrightarrow H^+_{(aq)} + HCO^-_{3(aq)} \\ &\leftrightarrow 2H^+_{(aq)} + CO_3^{2-}{}_{(aq)} \end{aligned} \quad (14)$$

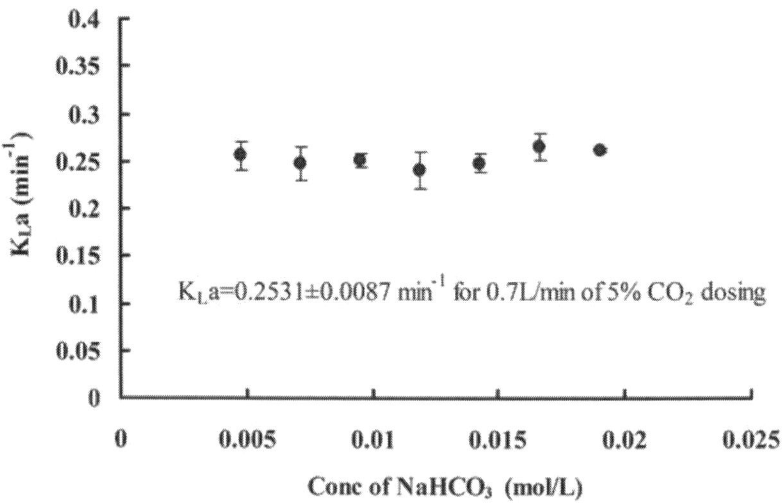

Figure 9. Changes in CO_2 mass transfer coefficients for different concentrations of $NaHCO_3$.

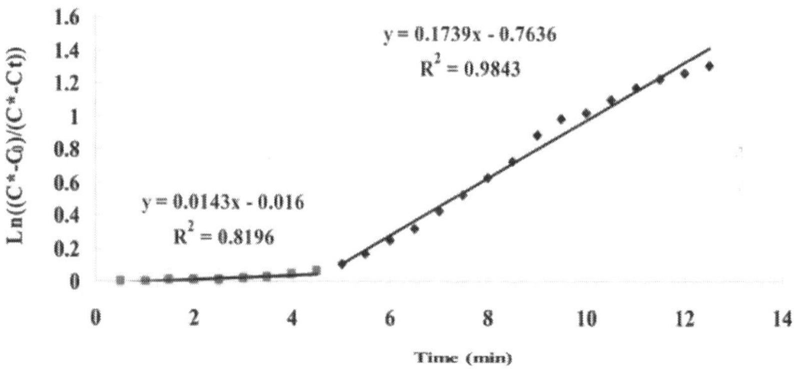

Figure 10. Typical plot for K_La estimation (for 0.7 L·min^{-1} dosing).

The slop of straight line indicates mass transfer coefficient K_La (min^{-1}).

In total, the CO_2 mass transfer coefficient in the D. salina culture with 0.7 L·min^{-1} of 5% CO_2 gas dosing was found to be 0.0164 ± 0.0046 min^{-1} at pH > 8.4 and 0.1776 ± 0.0064 min^{-1} at pH < 8.4. For clarification purpose, at pH > 8.4 the K_La for dissolved CO_2 is relatively low, but it dose not mean less CO_2 from gas supply has been transferred into liquid. Actually, most of the CO_2 been transferred into culture was converted into HCO_3^- and CO_3^{2-} when pH > 8.4. Therefore, when calculating the CO_2 capture efficiency in future, the changes in the amount of total carbon (C_T) should be considered rather than the amount of dissolved CO_2 when pH > 8.4. But if the pH is less than 8.4, the changes in the amount of total carbon almost come from the changes in dissolved CO_2, so it is fair to use the K_La of dissolved CO_2 for CO_2 capture efficiency estimation.

Comparing both CO_2 mass transfer (when pH < 8.4) under 0.7 $L \cdot min^{-1}$ of dosing for water containing $NaHCO_3$ and the culture medium including microalgae (with the same concentration of $NaHCO_3$), the K_La in water (0.2531 min^{-1}) was found to be greater than the one in the presence of D. salina (0.1776 min^{-1}). That may be because the cells in the medium increased its viscosity, which could have reduced the diffusivity of CO_2 from liquid film to liquid phase plus part of dissolved CO_2 could be consumed due to D. salina growth. Hence the rate of CO_2 diffusion into the culture was slowed down, whilst without D. salina present the CO_2 could diffuse much easier through the medium and without being consumed. Also, because of the changes in liquid properties, the CO_2 equilibrium concentration $[CO_2]^*$ was slightly smaller in the culture (0.0011 ± 0.0001 $mol \cdot L^{-1}$) than that in the $NaHCO_3$ medium (0.0017 ± 0.0001 $mol \cdot L^{-1}$).

4. CONCLUSIONS

For the same bubble generation method, enhancing the gas dosing flowrate can increase the mass transfer coefficient. For the same bubbling flowrate, reducing the bubble size can lead to an improvement on K_La as well. Compared with fine-bubble dosing, microbubbles dosing of 5% CO_2gas by using fluidic oscillator has been proved to enhance the K_La for both CO_2 dissolution and O_2removal by 30% - 100% across different flow rate. Despite K_La can be enhanced by either increasing the dosing flowrate (to be more accurate, flowrate/liquid volume ratio) or reducing the bubble size, increasing flow rate to achieve a higher K_La would also raise the amount of CO_2 being wasted (not dissolved) which would ultimately lower the CO_2 capture efficiency. Therefore, in order to achieve both higher CO_2 mass transfer rate and capture efficiency for the improvement of microalgal growth and CO_2 sequestration, reducing bubble size (e.g. using microbubbles) is more promising than increasing flow rate.

The K_La for CO_2 dissolution was not affected by the presence of $NaHCO_3$, and $NaHCO_3$ could be used to control the equilibrium pH of the medium without affecting the $[CO_2]^*$ and K_La. The pH region can also be altered depending on the particular strain of microalgae being cultured, as different algae prefer different pH. In the real algal culture, there appears to be two distinct stages in terms of K_La, divided by the pH of 8.4. When pH < 8.4, due to the changes in liquid properties and carbon system equilibrium relations, the K_La as well as $[CO_2]^*$ was found slightly reduced than the ones in the water.

Future works need to be done to test the effect of different percentage of CO_2 in the gas supply on mass transfer. And a mathematical model correlating mass transfer to bubble size, flow rate/liquid volume ratio and percentage of CO_2 in the gas supply etc. is expected to be established, which could facilitate the estimation of CO_2 dosing time for microalgae culture.

ACKNOWLEDGEMENTS

We acknowledge support for microbubble dynamics from EPSRC. WZ would like to thank the Royal Society for a Brain Mercer Innovation Award. DJG would like to acknowledge support from Carbon Trust.

REFERENCES

1. L. Xu, P. J. Weathers, X.-R. Xiong and C.-Z. Liu, "Microalgal Bioreactors: Challenges and Opportunities," Engineering in Life Sciences, Vol. 9, No. 3, 2009, pp. 178- 189.

2. A. Richmond, "CRC Handbook of Microalgal Mass Culture. Boca Raton," CRC Press, Florida, 1986.

3. A. P. Carvalho, L. A. Meireles and F. X. Malcata, "Microalgal Reactors: A Review of Enclosed Systems Design and Performance," Biotechnology Progress, Vol. 22, No. 6, 2006, pp. 1490-1506.

4. A. Richmond, "Handbook of Microalgal Culture: Biotechnology and Applied Phycology," Wiley-Blackwell, New York, 2008.

5. O. Pulz, "Photobioreactors: Production Systems for Phototrophic Microorganisms," Applied Microbiology and Biotechnology, Vol. 57, No. 3, 2001, pp. 287-293.

6. R. K. Gary, "The Concentration Dependence of the ΔS Term in the Gibbs Free Energy Function: Application to Reversible Reactions in Biochemistry," Journal of Chemical Education, Vol. 81, No. 11, 2004, pp. 1599-1604.

7. A. P. Carvalho and F. X. Malcata, "Transfer of Carbon Dioxide within Cultures of Microalgae: Plain Bubbling Versus Hollow-Fiber Modules," Biotechnology Progress, Vol. 17, No. 2, 2001, pp. 265-272.

8. D. R. Ketkar, R. Mallikarjunan and S. Venkatachalam, "Electroflotation of Quartz Fines," International Journal of Mineral Processing, Vol. 31, No. 1, 1991, pp. 127-138.

9. J. K. Edzwald, J. P. Walsh, G. S. Kaminski and H. J. Dunn, "Flocculation and Air Requirements for Dissolved Air Flotation," American Water Works Association, Vol. 84, No. 3, 1992, pp. 92-100.

10. S. E. Rijk and J. G. Blanken, "Bubble Size in Flotation Thickening," Water Research, Vol. 28, No. 2, 1994, pp. 465-473.

11. C. Tsouris, D. W. DePaoli, J. Q. Feng and T. C. Scott, "Experimental Investigation of Electrostatic Dispersion of Nonconductive Fluids into Conductive Fluids," Industrial and Engineering Chemistry Research, Vol. 34, No. 4, 1995, pp.1394-1403.

12. L. A. Féris and J. Rubio, "Dissolved Air Flotation (DAF) Performance at Low Saturation Pressures," Filtration & separation, Vol. 36, No. 9, 1999, pp. 61-65.

13. W. B. Zimmerman, V. Tesar, S. Butler and H. C. H. Bandalusena, "Microbubble Generation," Recent Patents on Engineering, Vol. 2, No. 1, 2008, pp. 1-8.

14. W. B. Zimmerman, B. N. Hewakanadamby, V. Tesar, H. C. H. Bandulasena and O. A. Omotowa, "On the Design and Simulation of Airlift Loop Bioreactor with Microbubble Generation by Fluidic Oscillation," Food and Bioproducts Processing, Vol. 87, No. 10, 2009, pp. 215-227.

15. W. B. Zimmerman, M. Zandi, H. C. H. Bandulasena, V. Tesař, D. J. Gilmour and K. Ying, "Design of an Airlift Loop Bioreactor and Pilot Scales Studies with Fluidic Oscillator Induced Microbubbles for Growth of a Microalgae Dunaliella salina," Applied Energy, Vo. 88, No. 10, 2011, pp. 3357-3369.

16. M. Y. Chisti, "Airlift Bioreactors," Elsevier Applied Science, London, 1989.

17. K. Ying, D. J. Gilmour, Y. Shi and W. B. Zimmerman, "Growth Enhancement of Dunaliella salina by Microbubble Induced Airlift Loop Bioreactor (ALB)—The Relation between Mass Transfer and Growth Rate," Journal of Biomaterials and Nanobiotechnology, No. 4, 2013, pp. 1-9.

18. R. Corfield, "The Carbon Cycle," In: C. Cockell, Ed., An Introduction to the Earth-Life System, Cambridge University Press, Cambridge, 2008, pp. 61-93.

19. A. Stemler, "Forms of Dissolved Carbon Dioxide Required for Photosystem II Activity in Chloroplast Membranes," Plant Physiology, Vol. 65, No. 6, 1980, pp. 1160-1165.

CHAPTER 6

Phosphane-Based Cyclodextrins as Mass Transfer Agents and Ligands for Aqueous Organometallic Catalysis

Sébastien Tilloy, Cécile Binkowski-Machut, Stéphane Menuel, Hervé Bricout and Eric Monflier

Unité de Catalyse et de Chimie du Solide (UCCS), CNRS, UMR 8181, Univ. Artois, Rue Jean Souvraz, SP 18, F-62307 Lens, France

ABSTRACT

The replacement of hazardous solvents and the utilization of catalytic processes are two key points of the green chemistry movement, so aqueous organometallic catalytic processes are of great interest in this context. Nevertheless, these processes require not only the use of water-soluble ligands such as phosphanes to solubilise the transition metals in water, but also the use of mass transfer agents to increase the solubility of organic substrates in water. In this context, phosphanes based on a cyclodextrin skeleton are an interesting alternative since these compounds can simultaneously act as mass transfer agents and as coordinating species towards transition metals. For twenty years, various cyclodextrin-functionalized phosphanes have been described in the literature. Nevertheless, while their coordinating properties towards transition metals and their catalytic properties were fully detailed, their mass transfer agent properties were much less discussed. As these mass transfer agent properties are directly linked to the availability of the cyclodextrin cavity, the aim of this review is to demonstrate that the nature of the reaction solvent and the nature of the linker between cyclodextrin and phosphorous moieties can deeply influence the recognition properties. In addition, the impact on the catalytic activity will be also discussed.

KEYWORDS

cyclodextrin; hydroformylation; hydrogenation; inclusion; phosphanes; self-inclusion; water-soluble

1. INTRODUCTION

Green chemistry is a movement whose goal is to develop more environmentally friendly methods for the chemical industry and thus to reduce its environmental impact. Replacement of hazardous solvents with water is one of the main points of this movement [1]. In this field, aqueous organometallic catalysis occupies a prominent place [2]. Indeed, the use of organic solvents is not required and the water-soluble organometallic catalyst can be easily separated from products by decantation of the aqueous and organic phases at the end of the reaction. Nevertheless, aqueous organometallic catalysis processes present two main drawbacks. First, mass transfer agents, such as cyclodextrins (CDs) [3], are necessary to increase the low water-solubility of organic substrates and secondly, water-soluble ligands, such as phosphanes [4], are also necessary to solubilise the metal in water. In this context, cyclodextrin-functionalized phosphanes are an interesting alterative since these compounds can simultaneously act as mass transfer agents and as coordinating species (Figure 1).\

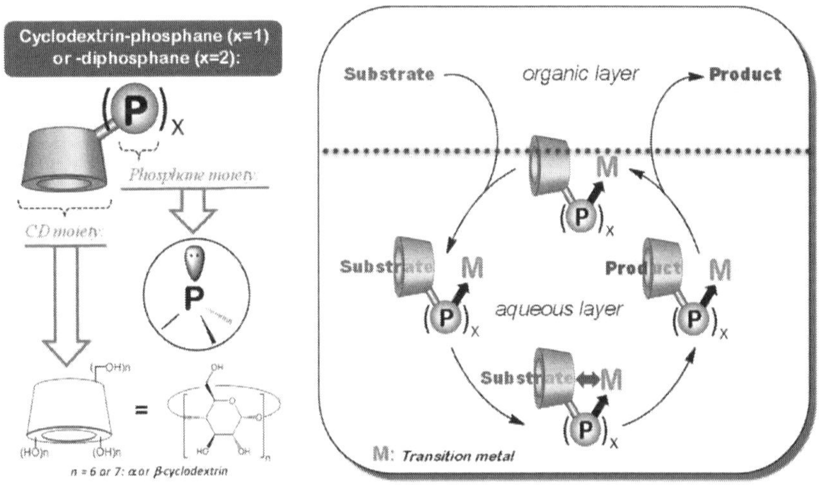

Figure 1. Principle of aqueous biphasic organometallic catalysis in the presence of cyclodextrin-functionalized phosphanes.

Indeed, CDs are water-soluble chemical receptors which are able to increase the water-solubility of organic substrates by inclusion into their hydrophobic cavity. In addition, the metal can be coordinated and rendered water-soluble thanks to the phosphane moiety covalently attached to the CD platform. For twenty years, various CD-mono, di, tri or tetra-phosphanes have been described in the literature [5,6,7,8,9,10,11,12,13,14,15,16, 17,18,19,20,21, 22,23,24,25,26,27], but only some CD-functionalized by one or two phosphorous moieties were exploited in aqueous organometallic catalytic processes [28,29,30, 31,32,33,34, 35,36]. Nevertheless, while their coordinating properties towards a transition metal and their catalytic properties were fully detailed, their mass transfer agent properties were much less discussed. As these mass transfer agent properties are directly linked to the recognition capacities, the aim of this review is to demonstrate that the nature of the reaction solvent and the nature of the linker between cyclodextrin and phosphorous moieties can deeply influence the availability of the CD cavity. In addition, the impact on the catalytic activity will be also discussed.

2. RESULTS AND DISCUSSION

2.1. Cyclodextrin-Diphosphanes

Historically, the first studies were realized on the CD-diphosphanes series. The first example of phosphanes covalently bond to a CD and used during an aqueous organometallic process was described in 1997 by the Reetz group [36]. More precisely, this group has proposed three different types of CD-diphosphanes (Figure 2; CD-diphosphanes **1–3**) in order to combine phase transfer catalysis and transition metal catalysis. To establish whether these phosphane-CDs functioned as supramolecular catalysts, an equimolar mixture of 1-decene and 4-phenyl-1-butene were subjected to competition experiments. This mixture was hydrogenated in DMF by rhodium complexes based on these CD-diphosphanes.

As a control experiment, a β-CD-free ligand was used instead of **1**, **2** or **3** and no substrate selectivity was achieved (Table 1, entry 1). In contrast, the use of CD-diphosphanes as ligands induced the preferential conversion of the phenyl-substituted alkene (Table 1, entries 2–5). This phenomenon was exacerbated when a DMF/H_2O mixture was used as solvent (entries 6–8). The authors postulated that a recognition step precedes the hydrogenation and furthermore that the phenyl group preferentially enters into the hydrophobic cavity of the β-CD framework. This observation was supported by the fact that the substrate selectivity drastically decreases when working in the presence of *p*-xylene which competes with substrates for the space in the CD cavity (entry 9). Nevertheless, no spectroscopic proof to support this hypothesis was furnished.

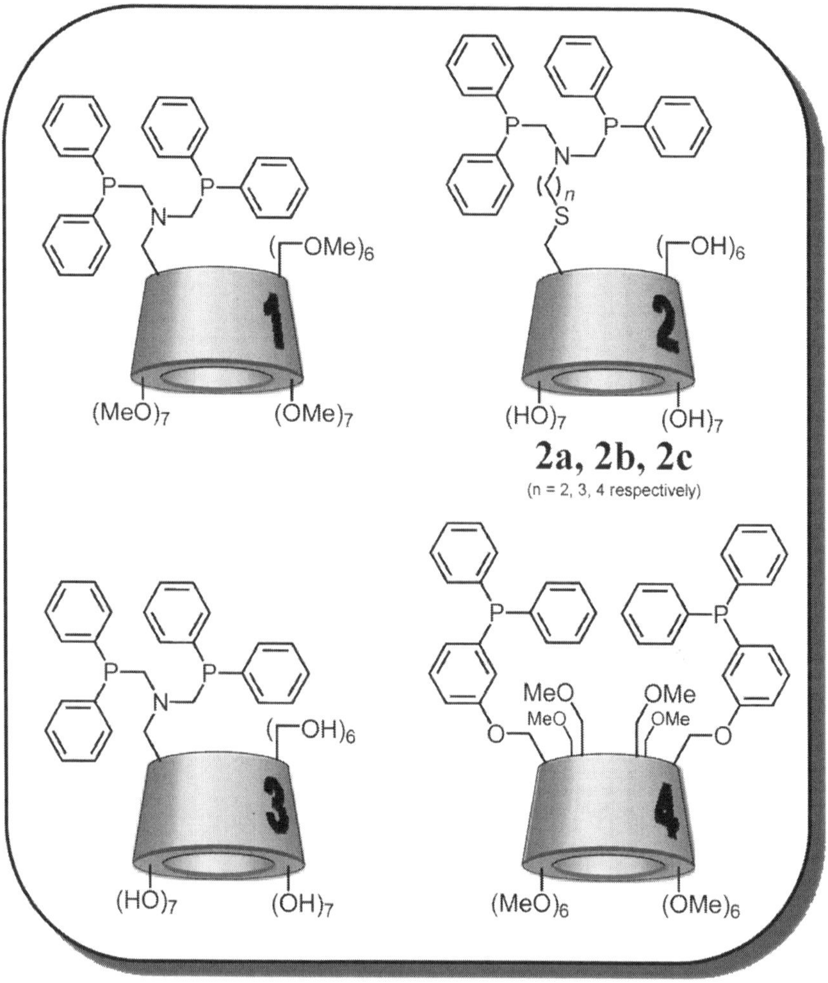

Figure 2. Schematic representation of cyclodextrin-diphosphanes.

Table 1. Substrate selectivity in hydrogenation.

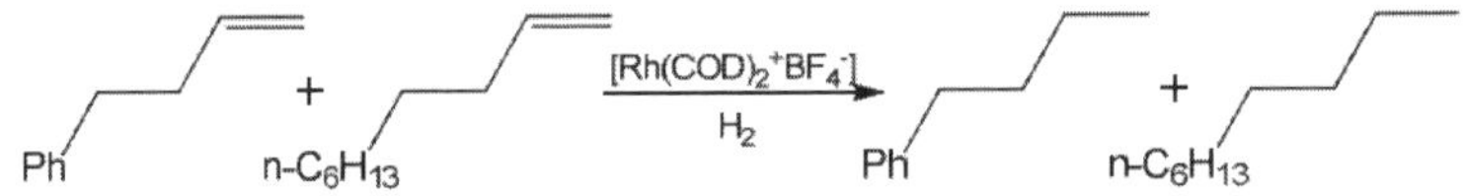

A B

Entry	Ligand	Solvent	A/B
1	$PhN(CH_2PPh_2)_2$	DMF	50/50
2	**1**	DMF	68/32
3	**2a**	DMF	74/26
4	**2b**	DMF	71/29
5	**2c**	DMF	66/34
6	**1**	DMF/H_2O (30/70)	82/18
7	**2a**	DMF/H_2O (30/70)	81/19
8	**3**	DMF/H_2O (30/70)	87/13
9 *	**3**	DMF/H_2O (30/70)	57/43

* In the presence of *p*-xylene.

Associated to a rhodium precursor, these β-CD-modified ligands were also tested in hydroformylation reactions in a two-phase system H_2O (with 30% DMF)/olefin (Scheme 1). These complexes are able to perform hydroformylation of 1-octene, (*E*)-3-hexene and allylbenbenzene. In the case of 1-octene, unexpectedly high catalytic activities were observed with a chemoselectivity for aldehydes superior to 99%. These systems are 150 times more active at 80 °C than the traditional system Rh/TPPTS (TPPTS: tris(*m*-sulfonatophenyl)phosphane trisodium salt) at 120 °C. High regioselectivity was also observed. For instance, the change from Rh/$PhN(CH_2PPh_2)_2$ in a one-phase system to Rh/**2a** in a two-phase system H_2O (with 30% DMF)/olefin results in an increase in the linear/branched aldehydes ratio from 62/38 to 76/24. The authors explained this enhanced selectivity by the participation of a host-guest complex between the substrate and the CD cavity as exposed in the Figure 3. It is important to underline that in the presence of toluene, which also enters the CD cavity and thus plays the role of a competitor, the regioselectivity was reduced to 65/35.

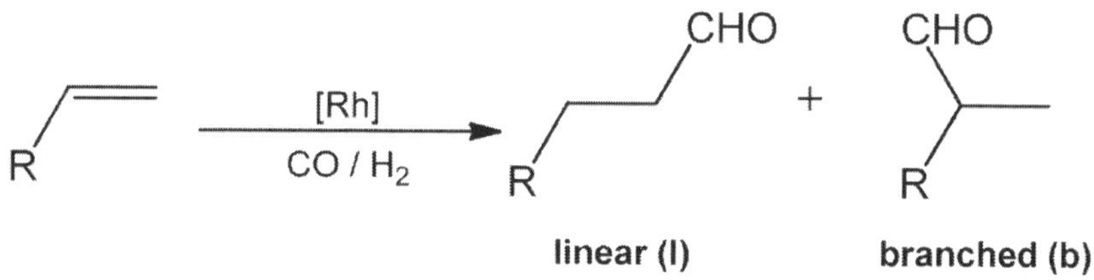

Scheme 1. Rhodium-catalysed hydroformylation of terminal olefin.

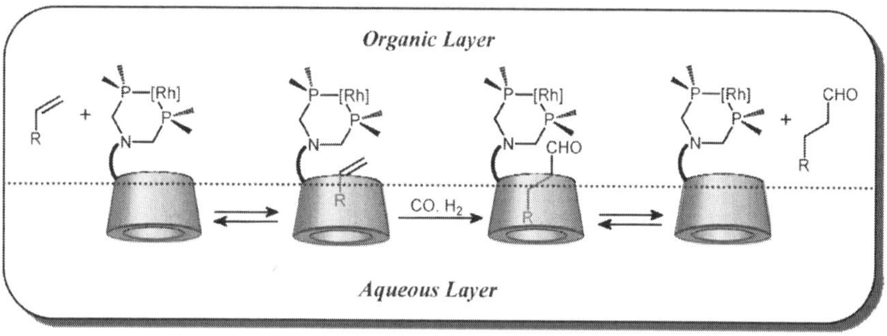

Figure 3. Mode of action of host-guest complexes in phase-transfer and rhodium catalysis proposed by Reetz.

During the reduction reaction of nitro-halogenated aromatic derivatives in a biphasic mixture of water/toluene, **1** proved also to be very efficient when associated with Pt and Rh-precursors to produce the corresponding aromatic amines [33]. The authors postulated that the catalytic system (Pt/**1** or Rh/**1**) acts as a phase transfer catalyst probably on the basis of host/guest phenomena as already described above. Oddly, the use of toluene as organic layer was not discussed. Indeed, in the preceding hydroformylation work, it was stated that the presence of toluene modified the catalytic activity since it is a competitor for the substrate.

The recovery and the recycling of the catalytic system during hydroformylation reactions were studied. The organic and aqueous phases were separated and unfortunately rhodium was detected by atomic absorption in the organic phase, showing a loss of the catalyst from the aqueous phase. Consequently, the reuse of the aqueous phase yielded a residual catalytic activity of approximately 50%. These last experiments showed that the ligands **1–3** were not able to effectively keep the rhodium in the aqueous phase. In addition, the presence of DMF which is used to facilitate the preparation of the catalyst, by increasing the solubility of the CD-phosphanes in water, also favoured the leaching of the metal into the organic phase. The presence of DMF has also a detrimental effect on the inclusion complexes. Indeed, DMF is a dissociating solvent which is well known to destabilize host-guest complexes [37]. Thus, in these preceding experiments, the presence of DMF disfavors not only the formation of stable inclusion complexes between the CD cavity and the substrate, but also leads to a loss of the catalyst in the organic layer.

Following these pioneering works, Matt and coworkers described the synthesis of a diphosphane based on permethylated α-cyclodextrin (Figure 2; CD-diphosphane **4**) [32]. This ligand is capable of complexing metals such as gold, silver, platinum, palladium and rhodium. The complex formed with rhodium has been tested in the hydroformylation reaction of 1-octene in biphasic aqueous medium (water/MeOH: 60/40). The conversion and chemoselectivity are greater than 99% and the l/b ratio is equal to 2.3. Unfortunately, as underlined by the authors, low water-solubility prevents "the catalyst" from acting as a supramolecular catalyst and leads to a significant loss of catalyst in the organic phase. Indeed, as in the case of DMF in the Reetz works, the presence of methanol, necessary as co-solvent to solubilize **4** in water, impedes the formation of stable inclusion complexes and leads to a loss of metal in the organic layer.

To summarize this part, these CD-diphosphanes **1–4** are poorly soluble in water forcing the use of a co-solvent with two major detrimental consequences: an increase in the loss of the catalyst in organic layer and a decrease in the CD/substrate recognition process. Since the presence of a co-solvent was necessary to increase the low water-solubility of these CD-diphosphanes leading to an adverse effect, CD-monophosphanes were synthetized. Indeed, the removal of one diphenylphosphino group was expected to reduce the compound hydrophobicity.

2.2. Cyclodextrin-Monophosphanes

The challenge was to synthetize CD-monophosphanes which are highly water-soluble and which possess highly recognition capacity in order to play both the role of mass transfer agent and coordinating species.

In 1998, Reetz and coworkers developed a monophosphane β-CD based on a native β-CD platform linked to a PPh_2 group by a thioethylene unit (Figure 4; CD-phosphane **5**). There was no information on the water-solubility of **5**, but the self-inclusion of one phenyl group of the PPh_2 moiety inside the CD cavity was shown by X-ray diffraction spectroscopy (Figure 5; CD-phosphane **5**) [38]. This ligand was evaluated in aqueous rhodium-catalyzed hydrogenation of olefins [34]. Unfortunately, hydrogenation occurred only under forcing conditions, leading to the formation of colloidal rhodium which then acted as a poor catalyst.

In this context, between 2010 and 2012, our team has reported the synthesis and the catalytic applications of three new water-soluble CD-phosphanes. These CD-phosphanes were built on a per- or randomly-methylated β-CD platform attached to a diphenylphosphino group (PPh_2). More precisely, the PPh_2 group has replaced one OMe group of permethylated β-CD (Figure 4; CD phosphane **6**) [31] or was bound to the β-CD skeleton by two different kinds of linkers: a *p*-phenylenoxy (Figure 4; CD phosphane **7**) [30] or a 4-methylene-1,2,3-triazolyl group (Figure 4; CD phosphane **8**) [29].

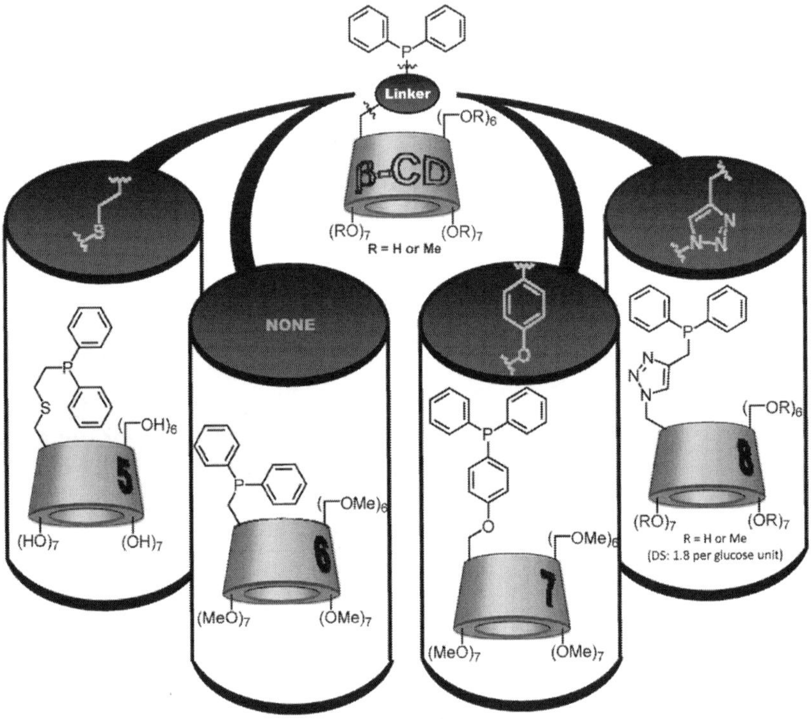

Figure 4. Schematic representations of cyclodextrin-monophosphanes.

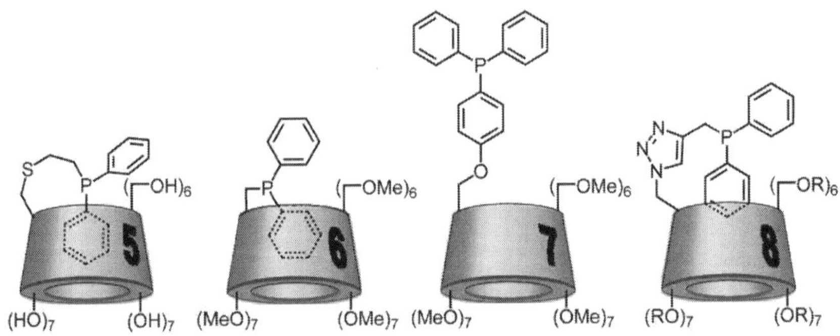

Figure 5. Schematic representation of preferential conformation of CD-phosphane **5–8**. For **8**, R= H or CH_3 with a substitution degree of 1.8.

The water-solubility of each CD-phosphane was equal to 3 mM, 0.01 mM and 15 mM at 20 °C for **6–8**, respectively. The difference between these values was firstly due to the nature of the CD platform. Indeed, the better water-solubility of **8** compared to **6** was due to the randomly methylated nature of its platform [39]. Secondly, the nature of the linker group between the CD and diphenylphosphino moieties directly impacted the positioning of the hydrophobic PPh_2 group regarding the cavity and thus the water-solubility. In fact, as observed above in the case of the CD-phosphane **5**, intramolecular complexation can occur according to the nature of the linker.

For the three CD-phosphanes **6**, **7** or **8**, the potential self-inclusion of the diphenylphosphino group was studied by 2D T-ROESY NMR experiments. Based on these data, schematic representations were gathered in the Figure 5. For **6**, a real self-inclusion phenomenon is highlighted where one phenyl group is included in the CD cavity whereas for **8**, this behaviour is less marked and the primary face of the CD is only capped by one of the two phenyl groups. In the case of **7**, the diphenylphosphino group is outside the cavity.

As the availability of the CD cavity is primordial during a phase transfer catalysis process, some inclusion experiments were performed in the presence of an external guest. Sodium 1-adamantanecarboxylate (ACNa) was selected because of its high affinity for the β-CD cavity. Indeed, the value of association constant for the β-CD/ACNa complex was equal to 30,000 M^{-1} [40].

The study of the dipolar interactions through 2D T-ROESY NMR experiments performed on a mixture**7**/ACNa in D_2O clearly revealed spatial proximity between the inner CD protons and guest protons proving the reality of an inclusion phenomenon. These observations show that **7** possess inclusion capacities. Attempts were also performed to include ACNa in the cavity of **6**. Interestingly, T-ROESY experiment showed dipolar contact not only between ACNa protons and CD protons, but also between phenyl protons and CD protons.

Thus, the guest is not able to fully expel the self-included aromatic ring from the cavity. This phenomenon seems to prove that the inclusion capacity of **6** is low. This assertion was supported by the determination of the value of association constant with ACNa. This constant was found equal to 265 M^{-1}. For comparison, the value of the association constant between ACNa and Trime-β-CD (permethylated β-CD) is equal to 10,700 M^{-1}. This value, 40 times higher than in the case of **6**, signifies that the self-inclusion complex is very strongly stabilized and the occupation of the cavity of **6** by the aromatic moiety will prevent the formation of stable inclusion complexes. 2D T-ROESY NMR experiments performed on a solution containing **8** and ACNa showed correlation peaks between the adamantyl protons and internal CD protons. In addition, the correlation between the phenyl and internal CD protons observed in the case of **8** alone has disappeared. These observations unambiguously proved that ACNa is able to fully expulse the partially included phenyl group. The value of association constant for **8**/ACNa was found to be very high since equal to 46,600 M^{-1}. This last result showed that although a capping phenomenon was observed, the CD cavity of**8** is able to deeply include a guest. Schematic representations of inclusion of ACNa in the cavity of **6**, **7** or **8**are summarized in Figure 6.

All these preceding results showed that the water-solubility and the inclusion capacity of these CD-phosphanes depends on the nature of the linker group between the CD and diphenylphosphino moieties, so the influence of this linker was evaluated during catalytic experiments.

Associated to a rhodium precursor, the three CD-phosphanes **6–8** were efficient in aqueous hydrogenation of 2-methyl-3-buten-2-ol and in aqueous hydroformylation of 4-acetoxystyrene and methyl 4-pentenoate. These experiments showed that **6–8** were valuable ligands for the transformation of various water-soluble substrates [29,30,31].

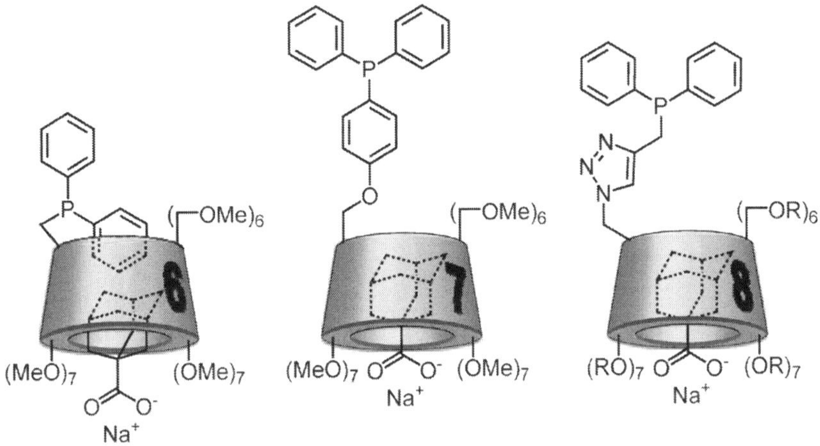

Figure 6. Schematic representation of the inclusion complexes between**6**, **7** or **8** and ACNa. For **8**, R = H or CH_3 with a substitution degree of 1.8.

The efficiency of these three CD-phosphanes as both ligands and mass transfer agent was explored in the case of 1-decene as an example of a water-insoluble substrate [41]. When rhodium complexes of CD-phosphanes **6** and **7** were tested in the hydroformylation reaction of 1-decene in biphasic aqueous medium, the conversion and chemoselectivity were greater than 97% with a l/b ratio equal to 2.8. Unfortunately, CD-phosphanes **6** or **7** are partially soluble in organic solvents. For example, the partition coefficient of **6**between water and heptane is equal to 0.48, that is, 68% of **6** stays in water and 32% moves to the organic layer. This behaviour is not in accordance with the specifications required for an application in biphasic aqueous organometallic catalytic processes. Indeed, a loss of the catalyst was observed in the organic layer thus impeding the recycling. When CD-phosphane **8** was used as ligand, the conversion did not exceed 5% after 6 h of reaction. This low conversion suggested that this ligand allowed immobilizing the rhodium in the aqueous layer. Unfortunately, it seems that 1-decene was not able to inverse the capping phenomenon by expulsing the phenyl group of the CD cavity. This behaviour prevents the transport of the olefin into the aqueous layer. To summarize this second part, the described CD-monophosphanes are not able to efficiently play both the role of mass transfer agent and catalytic species because of partial solubility in organic solvent or self-inclusion phenomena.

3. CONCLUSIONS

In this paper, we showed that it is not easy to synthetize CD-phosphanes able to act both as ligands and as phase transfer catalysts. Indeed, these CD-phosphanes have to be highly water-soluble to maintain the catalyst in aqueous layer and their hydrophobic cavity has to be available to include the substrate. The ideal CD-phosphane is expected to coordinate the metal center and to simultaneously form an inclusion complex with the substrate in order to achieve specific catalytic activity and selectivity. Up until 2012, no described CD-phosphane fulfills these specifications and the design of this ideal CD-phosphane remains a great challenge.

REFERENCES

1. Anastas, P.T.; Warner, J.C. *Green Chemistry: Theory and Practice*; Oxford University Press: New York, NY, USA, 1998.

2. Cornils, B.; Herrmann, W.A. *Aqueous-Phase Organometallic Catalysis*; Cornils, B., Herrmann, W.A., Eds.; Wiley-VCH: Weinheim, Germany, 2004; p. 1.

3. Hapiot, F.; Ponchel, A.; Tilloy, S.; Monflier, E. Cyclodextrins and their applications in aqueous-phase metal-catalyzed reactions. *C.R. Chim.* **2011**, *14*, 149–166.

4. Shaughnessy, K.H. Hydrophilic ligands and their application in aqueous-phase metal-catalyzed reactions. *Chem. Rev.* **2009**, *109*, 643–710.

5. Gramage-Doria, R.; Rodriguez-Lucena, D.; Armspach, D.; Egloff, C.; Jouffroy, M.; Matt, D.; Toupet, L. A cavity-shaped diphosphane displaying oschelating behaviour. *Angew. Chem. Int. Ed.* **2011**, *50*, 1554–1559.

6. Zaborova, E.; Deschamp, J.; Guieu, S.; Blériot, Y.; Poli, G.; Ménand, M.; Madec, D.; Prestat, G.; Sollogoub, M. Cavitand supported tetraphosphine: Cyclodextrin offers a useful platform for Suzuki-Miyaura cross-coupling. *Chem. Commun.* **2011**, *47*, 9206–9208.

7. Gramage-Doria, R.; Rodriguez-Lucena, D.; Armspach, D.; Egloff, C.; Jouffroy, M.; Matt, D.; Toupet, L. Regioselective double capping of cyclodextrin scaffolds. *Chem. Eur. J.* **2011**, *17*, 3911–3921.

8. Guieu, S.; Zaborova, E.; Blériot, Y.; Poli, G.; Jutand, A.; Madec, D.; Prestat, G.; Sollogoub, M. Can hetero-polysubstituted cyclodextrins be considered as inherently chiral concave molecules? *Angew. Chem. Int. Ed.* **2010**, *49*, 2314–2318.

9. Armspach, D.; Matt, D.; Toupet, L. Self-mediated, stereoselective oxidation of thia-capped cyclodextrins. *Angew. Chem. Int. Ed.* **2009**, *48*, 4555–4558.

10. Poorters, L.; Armspach, D.; Matt, D.; Toupet, L.; Jones, P.G. A metallocavitand functioning as a container for anions: formation of noncovalent linear assemblies mediated by a cyclodextrin-entrapped NO_3–ion. *Angew. Chem. Int. Ed.* **2007**, *46*, 2663–2665.

11. Poorters, L.; Armspach, D.; Matt, D.; Toupet, L.; Choua, S.; Turek, P. Synthesis and properties of TRANSDIP, A rigid chelator built upon a cyclodextrin cavity: Is TRANSDIP an authentic trans-spanning ligand? *Chem. Eur. J.* **2007**, *13*, 9448–9461.

12. Poorters, L.; Armspach, D.; Matt, D.; Toupet, L. α-TEPHOS: A cyclodextrin-derived tetraphosphine for multiple metal binding. *Dalton Trans.* **2007**, *7*, 3195–3202.

13. Hapiot, F.; Tilloy, S.; Monflier, E. Cyclodextrins as Supramolecular Hosts for Organometallic Complexes. *Chem. Rev.* **2006**, *106*, 767–781.

14. Engeldinger, E.; Poorters, L.; Armspach, D.; Matt, D.; Toupet, L. Diastereospecific synthesis of phosphinidene-capped cyclodextrins leading to "introverted" ligands. *Chem. Commun.* **2004**, *21*, 634–635.

15. Engeldinger, E.; Armspach, D.; Matt, D.; Jones, P.G. Cyclodextrin phosphanes as first and second coordination Sphere Cavitands. *Chem. Eur. J.* **2003**, *9*, 3091–3105.

16. Poorters, L.; Armspach, D.; Matt, D. Selective tetrafunctionalisation of alpha-cyclodextrin using the supertrityl protecting group. Synthesis of the first C2-symmetric tetraphosphine based on a cavitand (alpha-TEPHOS). *Eur. J. Org. Chem.* **2003**, *2003*, 1377–1381.

17. Engeldinger, E.; Armspach, D.; Matt, D. Capped cyclodextrins. *Chem. Rev.* **2003**, *103*, 4147–4173.

18. Engeldinger, E.; Armspach, D.; Matt, D.; Toupet, L.; Wesolek, M. . Synthesis of large chelate rings with diphosphites built on a cyclodextrin scaffold. Unexpected formation of 1,2-phenylene-capped α-cyclodextrins. *C.R. Chim.* **2002**, *5*, 359–372.

19. Reetz, M.T.; Kostas, I.D.; Waldvogel, S.R. Synthesis of a gold(I) complex with a (thio)phosphine-modified β-cyclodextrin. *Inorg. Chem. Commun.* **2002**, *5*, 252–254.

20. Engeldinger, E.; Armspach, D.; Matt, D.; Jones, P.G.; Welter, R. A cyclodextrin diphosphane as a first and second coordination sphere cavitand: evidence for weak C-H···Cl-M hydrogen bonds

within metal-capped cavities. *Angew. Chem. Int. Ed.* **2002**, *41*, 2593–2596.

21. Wong, Y.T.; Yang, C.; Ying, K.C.; Jia, G. Synthesis of a novel beta-cyclodextrin-functionalized diphosphine ligand and its catalytic properties for asymmetric hydrogenation. *Organometallics* **2002**,*21*, 1782–1787.

22. Yang, C.; Cheung, Y.K.; Yao, J.; Wong, Y.T.; Jia, G. Palladium and platinum complexes with a beta-cyclodextrin-functionalized phosphine ligand. *Organometallics* **2001**, *20*, 424–429.

23. Yang, C.; Wong, Y.T.; Li, Z.; Krepinsky, J.J.; Jia, G. Synthesis of beta-cyclodextrin-functionalized (2S,4S)-(-)-4-(diphenylphosphino)-2-(diphenylphosphinomethyl) ligands and their rhodium and platinum complexes. *Organometallics* **2001**, *20*, 5220–5224.

24. Engeldinger, E.; Armspach, D.; Matt, D. Cyclodextrin cavities as probes for ligand-exchange processes. *Angew. Chem. Int. Ed.* **2001**, *40*, 2526–2529.

25. Deshpande, R.M.; Fukuoka, A.; Ichikawa, M. Novel phosphinite capped cyclodextrin-rhodium catalysts in substrate selective hydroformylation. *Chem. Lett.* **1999**, *1*, 13–14.

26. Sawamura, M.; Kitayama, K.; Ito, Y. Synthesis and properties of a new chiral diphosphine ligand bearing a cyclodextrin-based molecular recognition site and its palladium(II) complex. *Tetrahedron: Asymmetry* **1993**, *4*, 1829–1832.

27. Reetz, M.T.; Rudolph, J. Synthesis of a phosphine-modified cyclodextrin and its rhodium complex.*Tetrahedron: Asymmetry* **1993**, *4*, 2405–2406.

28. Hapiot, F.; Bricout, H.; Tilloy, S.; Monflier, E. Functionalized Cyclodextrins as First and Second Coordination Sphere Ligands for Aqueous Organometallic Catalysis. *Eur. J. Inorg. Chem.* **2012**, *2002*, 1571–1578.

29. Tran, D.N.; Legrand, F.X.; Menuel, S.; Bricout, H.; Tilloy, S.; Monflier, E. Cyclodextrin-phosphane possessing a guest-tunable conformation for aqueous rhodium-catalyzed hydroformylation. *Chem. Commun.* **2012**, *48*, 753–755.

30. Legrand, F.X.; Six, N.; Slomianny, C.; Bricout, H.; Tilloy, S.; Monflier, E. Synthesis, Rhodium complexes and catalytic applications of a new water-soluble triphenylphosphane-modified β-cyclodextrin. *Adv. Synth. Catal.* **2011**, *353*, 1325–1334.

31. Machut-Binkowski, C.; Legrand, F.X.; Azaroual, N.; Tilloy, S.; Monflier, E. New phosphane based on a β-cyclodextrin exhibiting a solvent-tunable conformation and its catalytic properties. *Chem. Eur. J.***2010**, *16*, 10195–10201.

32. Armspach, D.; Matt, D. Metal-capped β-cyclodextrins: The crowing of the oligosaccharide torus with precious metals. *Chem. Commun.* **1999**, *12*, 1073–1074.

33. Reetz, M.T.; Frömbgen, C. Chemoselective reduction of halo-nitro aromatic compounds by β-cyclodextrin-modified transition metal catalysts in a biphasic system. *Synthesis* **1999**, *1999*, 1555–1557.

34. Reetz, M.T. Supramolecular transition metal catalysts in two-phase systems. *Catal. Today* **1998**, *42*, 399–411.

35. Reetz, M.T. New supramolecular transition metal catalysis. *J. Heterocycl. Chem.* **1998**, *35*, 1065–1073.

36. Reetz, M.T.; Waldvogel, S.R. Diphosphanes as ligands for supramolecular rhodium catalysts. *Angew. Chem. Int. Ed. Engl.* **1997**, *36*, 865–867.

37. Feliciano, C.E. Quiñones, E. The association of 4-(*N,N*-dimethylamino)benzonitrile and beta-cyclodextrin in dimethyl sulfoxide and N,N-dimethylformamide. *J. Photochem. Photobiol. A Chem.***1999**, *120*, 23–28.

38. Reetz, M.T.; Rudolph, J.; Goddard, R. Diastereotopic group recognition in the solid state- A unique intramolecular β-cyclodextrin inclusion complex. *Can. J. Chem.* **2001**, *79*, 1806–1811.

39. Loftsson, T.; Jahro, P.; Masson, M.; Jarvinen, T. Cyclodextrins in drug delivery. *Expert Opin. Drug Deliv.* **2005**, *2*, 335–351.

40. Rekharsky, M.V.; Inoue, Y. Complexation thermodynamics of cyclodextrins. *Chem. Rev.* **1998**, *98*, 1875–1910.

41. Legrand, F.X. New mass transfer agents and ligands based on cyclodextrin for aqueous organometallic catalysis. Ph.D. Thesis, University of Artois, Lens, France, 23 November 2010.

CHAPTER 7

Numerical Solution of MHD Boundary Layer Flow of Non-Newtonian Casson Fluid on a Moving Wedge with Heat and Mass Transfer and Induced Magnetic Field

Nabil T. El-Dabe[1], Ahmed Y. Ghaly[1], Raafat R. Rizkallah[1], Karem M. Ewis[2], Ameen S. Al-Bareda[1]

[1]Department of Mathematics, Faculty of Education, Ain Shams University, Cairo, Egypt

[2]Department of Engineering Mathematics and Physics, Faculty of Engineering, El-Fayoum University, El-Fayoum, Egypt

ABSTRACT

The paper investigates the numerical solution of the magnetohydrodynamics (MHD) boundary layer flow of non-Newtonian Casson fluid on a moving wedge with heat and mass transfer. The effects of thermal diffusion and diffusion thermo with induced magnetic field are taken in consideration. The governing partial differential equations are transformed into nonlinear ordinary differential equations by applying the similarity transformation and solved numerically by using finite difference method (FDM). The effects of various governing parameters, on the velocity, temperature and concentration are displayed through graphs and discussed numerically. In order to verify the accuracy of the present results, we have compared these results with the analytical solutions by using the differential transform method (DTM). It is observed that this approximate numerical solution is in good agreement with the analytical solution. Furthermore, comparisons of the pre- sent results with previously published work show that the present results have high accuracy.

KEYWORDS

Casson Fluid, Induced Magnetic Field, Boundary Layer, Moving Wedge, Radiation, Soret and Dufour Effects

1. INTRODUCTION

The induced magnetic field is arisen due to a strong magnetic field. The problems with induced magnetic field play an important role in a number of industrial applications such as fiber or granular insulation, liquid-metals, electrolytes, ionized gases as well as the geothermal systems. The study of the boundary layer flow under the influence of a magnetic field with the induced magnetic field was considered by a few authors. For example, Raptis and Perdikis [1] studied the MHD free convection boundary layer flow past an infinite vertical porous plate. Very little attention has been paid to the boundary layer flow and heat transfer taking the effect of the induced magnetic field and thermal radiation into consideration. For example, Kumari et al. [2] considered the MHD flow and heat transfer over a stretching surface by considering the effect of the induced magnetic field. On the other hand, Takhar et al. [3] investigated the unsteady free convection flow at the stagnation point in the presence of a magnetic field. Recently Ali et al. [4] studied MHD stagnation-point flow and heat transfer towards stretching sheet with induced magnetic field. Khamisah Jafar et al. [5] studied viscous dissipation and radiation effects in MHD stagnation point flow towards a stretching sheet with induced magnetic field. The thermal reversal in MHD stagnation point flow towards a stretching sheet with induced magnetic field and viscous dissipation effects have been investigated by Kashif Ali and Muhammad Ashraf [6] . The MHD mixed convection boundary layer flow toward a stagnation point on a vertical surface with induced magnetic field was studied by Ali et al. [7] . Khamisah Jafar et al. [8] discussed the MHD boundary layer flow due to a moving wedge in a parallel stream with the induced magnetic field.

The analysis of non-Newtonian fluids has many applications in industrials such as ground water hydrology, petroleum reservoir, nuclear waste disposal, geothermal energy production, transpiration cooling, design of solid matrix heat exchange and paced bed chemical catalytic reactors. Also, there are some models of non-Newtonian fluids such as Casson model which concerned one of the most important applications of theoretical fluid mechanics to problems arising in physiology, mainly in describing the flow of the blood. Casson [9] proposed a model to describe the flow curves of suspensions of pigments in lithographic varnishes used for the preparation of printing inks. Eldabe et al. studied some problems of non-Newtonian fluids such as ([10] - [14]).

In the present work, we extend and generalize the work of [8] to include the boundary layer motion of non- Newtonian fluid obeying Casson model with heat and mass transfer in the presence of radiation and mixed diffusion. The main objective of the present work is to obtain the numerical solutions of the MHD boundary layer flow of Casson fluid on a moving wedge with heat and mass transfer. The effects of thermal diffusion and diffusion thermo with induced magnetic field are taken into consideration. The governing equations of motion with the boundary conditions are solved numerically by using (FDM) and analytically by using (DTM). The solutions are obtained as a function of the physical parameters of the problem; then the effects of these various parameters of the problem on these solutions have been computed and discussed in detail and illustrated through some figures. The special cases of our problem are compared with the previous published works. This showed that there is a good agreement between them.

2. MATHEMATICAL FORMULATION

Let us consider a steady, laminar, hydromagnetic coupled heat and mass transfer by mixed convection flow on a wedge plate in a parallel free stream with a variable induced magnetic field applied parallel to the wedge walls outside the boundary layer. The equations governing the steady motion of an incompressible non-Newtonian electrically conducting fluid in presence of a magnetic field are:

$$\nabla \cdot V = 0, \ \nabla \cdot H = 0 \tag{1}$$

$$(V \cdot \nabla)V - \frac{\mu_0}{4\pi\rho}(H \cdot \nabla)H = -\frac{1}{\rho}\nabla P + \frac{1}{\rho}\nabla \cdot \tau, \tag{2}$$

$$\nabla \times (V \times H) + \mu_e \nabla^2 H = 0, \tag{3}$$

$$\rho c_p (V \cdot \nabla)T = k\nabla^2 T + \tau \cdot (\nabla V) - \frac{\partial q_r}{\partial y} + \frac{\rho D_m k_T}{c_s}\nabla^2 C \tag{4}$$

$$(V \cdot \nabla)C = D_m \nabla^2 C + \frac{D_m k_T}{T_m}\nabla^2 T \tag{5}$$

where V is the fluid velocity vector, H is the induced magnetic field vector, $P = p + \frac{\mu|H|^2}{8\pi}$ is the magnetohy-drodynamic pressure, p is the fluid pressure, μ_0 is the magnetic permeability, τ is the stress tensor, σ is the electrical conductivity, T is the temperature, k is the thermal conductivity, c_p is the specific heat at constant pressure, ρ is the density of the fluid, $\mu_e = \frac{1}{4\pi\sigma}$ is the magnetic diffusivity, q_r is the radiative heat flux, C is the species concentration, C_s is the concentration susceptibility, k_T is thermal diffusion ratio, T_m is mean fluid temperature and is the coefficient of chemical molecular diffusivity.

The rheological equation of state for an isotropic flow of a Casson fluid can be expressed as [9] :

$$\tau_{ij} = \begin{cases} 2\left(\mu_B + \frac{p_y}{\sqrt{2\pi}}\right)e_{ij}, & \pi > \pi_c \\ 2\left(\mu_B + \frac{p_y}{\sqrt{2\pi}}\right)e_{ij}, & \pi < \pi_c \end{cases} \tag{6}$$

In the above equation $\pi = e_{ij}e_{ij}$ and denotes the $(i,j)^{th}$ component of the deformation rate, π is the product of the component of deformation rate with itself, π_c is the a critical value of this product based on the non-Newtonian model, μ_B is the plastic dynamic viscosity of the non-Newtonian fluid and is the yield stress of the fluid.

We take the Cartesian coordinates x measured along the surface of the wedge and y normal to it, respectively. If (u,v) and (H_1,H_2) are the velocity and magnetic components in (x, y) directions, respectively, subject to the boundary layer approximations, Equations (5) and (6) for the problem under consideration can be reduced to:

$$\frac{\partial u}{\partial x}+\frac{\partial v}{\partial y}=0, \tag{7}$$

$$\frac{\partial H_1}{\partial x}+\frac{\partial H_2}{\partial y}=0, \tag{8}$$

$$u\frac{\partial u}{\partial x}+v\frac{\partial u}{\partial y}-\frac{\mu_0}{4\pi\rho}\left(H_1\frac{\partial H_1}{\partial x}+H_2\frac{\partial H_1}{\partial y}\right)=U_e\frac{dU_e}{dx}-\frac{\mu_0 H_e}{4\pi\rho}\frac{dH_e}{dx}+v\left(1+\frac{1}{\beta}\right)\frac{\partial^2 u}{\partial y^2}, \tag{9}$$

$$u\frac{\partial H_1}{\partial x}+v\frac{\partial H_1}{\partial y}-H_1\frac{\partial u}{\partial x}-H_2\frac{\partial u}{\partial y}=\mu_e\frac{\partial^2 H_1}{\partial y^2} \tag{10}$$

$$u\frac{\partial T}{\partial x}+v\frac{\partial T}{\partial y}=\alpha\frac{\partial^2 T}{\partial y^2}+\frac{v}{c_p}\left(1+\frac{1}{\beta}\right)\left(\frac{\partial u}{\partial y}\right)^2-\frac{1}{\rho c_p}\left(\frac{\partial q_r}{\partial y}\right)+\frac{D_m k_T}{c_s c_p}\left(\frac{\partial^2 C}{\partial y^2}\right) \tag{11}$$

$$u\frac{\partial C}{\partial x}+v\frac{\partial C}{\partial y}=D_m\frac{\partial^2 C}{\partial y^2}+\frac{D_m k_T}{T_m}\left(\frac{\partial^2 T}{\partial y^2}\right) \tag{12}$$

The boundary conditions are as follows:

$$\begin{gathered} v=0,\ u=u_w(x)=U_w x^m,\ H_1=H_2=0,\ T=T_w,\ C=C_w,\ \text{at } y=0 \\ u=U_e(x)=U_\infty x^m,\ H_1=H_e(x)=H_0 x^m, T=T_\infty, C=C_\infty\ \text{ as } y\to\infty \end{gathered} \tag{13}$$

where $\beta = \mu_B \sqrt{2\pi_c / p_y}$ is the non-Newtonian Casson parameter, v is the kinematic viscosity, $U_e(x)$ and $H_e(x)$ are the $\alpha = k/\rho c_p$ x-velocity and magnetic field at the edge of the boundary layer, respectively and is the thermal diffusivity.

We assume here that and $H_e(x) = H_0 x^m$, is the constant velocity at the outer edge of the boundary layer and H_0 is the value of at x = 0. Further, m is also a constant, which varies in the range, U_w is a positive or a negative constant [8]

By using the Rosseland approximation, the radiative heat flux q_r is given by given by

$$q_r = -\frac{4\sigma^*}{3k^*}\frac{\partial T^4}{\partial y} \tag{14}$$

where σ^* is the Stefan-Boltzmann constant and k^* is the mean absorption coefficient. It should be noted that by using the Rosseland approximation, the present analysis is limited to optically thick fluids. If temperature differences within the flow are sufficiently small, then Equation (14) can be linearized by expanding T^4 into the Taylor series abo T_∞, which after neglecting higher order terms takes the form

$$T^4 \approx 4T_\infty^3 T - 3T_\infty^4 \tag{15}$$

In view of Equations (14) and (15), Equation (11) reduces to

$$u\frac{\partial T}{\partial x} + v\frac{\partial T}{\partial y} = \alpha\left(1 + \frac{16\sigma^* T^3_\infty}{3kk^*}\right)\frac{\partial^2 T}{\partial y^2} + \frac{v}{c_p}\left(1 + \frac{1}{\beta}\right)\left(\frac{\partial u}{\partial y}\right)^2 + \frac{D_m k_T}{c_s c_p}\left(\frac{\partial^2 C}{\partial y^2}\right) \tag{16}$$

3. METHOD OF SOLUTIONS

To solve system of Equations (7)-(13), we will consider the following similarity transformations:

$$\eta=\sqrt{\frac{(m+1)U_e(x)}{2vx}}y,\ \psi=\sqrt{\frac{2vxU_e(x)}{(m+1)}}f(\eta),\ \varphi=H_0\sqrt{\frac{2vxx^m}{(m+1)U_\infty}}g(\eta),$$

$$\theta=\left(\frac{T-T_\infty}{T_w-T_\infty}\right),\ \varnothing=\left(\frac{C-C_\infty}{C_w-C_\infty}\right)$$

(17)

where η is the independent similarity variable, $f(\eta)$ the dimensionless stream function, $g(\eta)$ the dimensionless Induced magnetic field, $\theta(\eta)$ the dimensionless temperature and $\varphi(\eta)$ the dimensionless concentration. Further, ψ is the stream function which is defined in the usual way as and and φ is the magnetic stream function which is defined in the usual way as and By using Equation (17) in the Equations (9)-(12) and Equation (16), we get the following ordinary differential equations in dimensionless form

$$\left(1+\frac{1}{\beta}\right)f'''+ff''+\gamma\left(1-f'^2\right)-M\left[gg''+\gamma\left(1-g'^2\right)\right]=0$$

(18)

$$\lambda g'''+fg''-f''g=0 \qquad (19)$$

$$\left(1+\frac{4}{3}R_a\right)\theta''+P_rf\theta'+P_rE\left(1+\frac{1}{\beta}\right)f''^2+P_rD_u\varnothing''=0$$

(20)

$$\varnothing'' + S_c f \varnothing' + S_c S_0 \theta'' = 0 \tag{21}$$

where $\gamma = 2m/(m+1)$ is the wedge parameter, $M = \mu H_0^2 / 4\pi\rho U_\infty^2$ is the magnetic parameter, $\lambda = 1/4\pi v\sigma$ is the reciprocal magnetic Prandtl number, $E = U_e^2(x)/c_p(T_w - T_\infty)$ is the Eckert number, $P_r = v/\alpha$ is the Prandtl number, $R_a = 4\sigma T_\infty^3 / kk^*$ is the radiation parameter, $S_c = v/D_m$ is the Schmidt number, $D_u = D_m k_T (C_w - C_\infty)/vc_s c_p (T_w - T_\infty)$ is the Dufour number and $S_0 = D_m k_T (T_w - T_\infty)/vT_m (C_w - C_\infty)$ is the Soret number.

Also, the subjected boundary conditions Equation (13) will now take the form:

$$\begin{gathered} f(0)=0,\ f'(0)=\delta,\ g(0)=0,\ g'(0)=0,\ \theta(0)=1,\ \varnothing(0)=1, \\ f'(\infty)\to 1,\ g'(\infty)\to 1,\ \theta(\infty)\to 0,\ \varnothing(\infty)\to 0 \end{gathered} \tag{22}$$

where $\delta = U_w / U_\infty$ is the moving parameter.

3.1. The Differential Transform Method (DTM)

The differential transformation of an analytical function $f(t)$ for one variable is defined as [15] .

$$F(k) = \frac{1}{k!}\left[\frac{d^k f(t)}{dt^k}\right]_{t=t_0} \tag{23}$$

where $f(t)$ is the original function and $F(k)$ is the transformed function. The differential inverse transformation of $F(k)$ is defined as:

$$f(t)=\sum_{k=0}^{\infty}F(k)(t-t_0)^k \tag{24}$$

Combining Equations (23) and (24), we obtain

$$f(t)=\sum_{k=0}^{\infty}\frac{(t-t_0)^k}{k!}\left[\frac{\mathrm{d}^k f(t)}{\mathrm{d}t^k}\right]_{t=t_0} \tag{25}$$

From Equations (23)-(25), it can be seen that the differential transformation method is derived from Taylor's series expansion, but the method does not calculate the derivatives representatively. However, the relative derivatives are calculated by an iterative way which is described by the transformed equations of the original function. For implementation purposes, the function $f(t)$ is expressed by a finite series and Equation (24) can be written as

$$f(t)\approx\sum_{k=0}^{N}F(k)(t-t_0)^k \tag{26}$$

By Equation (23), the following theorems can be deduced:

Theorem 3.1.1 If $f(t)=u(t)\pm v(t)$ then $F(k)=U(k)\pm V(k)$.

Theorem 3.1.2 If $f(t)=\alpha u(t)$ then $F(k)=\alpha U(k)$.

Theorem 3.1.3 If $f(t)=t^m$ then $F(k)=\delta(k-m)=\begin{cases}1 & k=m\\ 0 & \text{otherwise}\end{cases}$.

Theorem 3.1.4 If $f(t)=\frac{\mathrm{d}u(t)}{\mathrm{d}t}$ then $F(k)=(k+1)U(k+1)$.

Theorem 3.1.5 If $f(t)=\frac{\mathrm{d}^n u(t)}{\mathrm{d}t^n}$ then $F(k)=\frac{(k+n)!}{k!}U(k+n)$.

Theorem 3.1.6 If $f(t)=u(t)\cdot v(t)$ then $F(k)=\sum_{r=0}^{k}U(r)V(k-r)$.

Theorem 3.1.7

If $f(t)=\frac{du(t)}{dt}\frac{du(t)}{dt}$ then $F(k)=\sum_{r=0}^{k}(r+1)(k-r+1)U(r+1)U(k-r+1)$.

3.2. Basic Concepts of the Multi-Step Differential Transform Method (MDTM)

When the DTM is used for solving differential equations with the boundary condition at infinity or problems that have highly non-linear behavior, the obtained results were found to be incorrect (when the boundary-layer variable go to infinity, the obtained series solutions are divergent). Besides that, power series are not useful for large values of the independent variable.

To overcome this shortcoming, the MDTM that has been developed for the analytical solution of the differential equations is presented in this section. For this purpose, the following non-linear initial-value problem is considered:

$$u\left(t,f,f',\cdots,f^{(p)}\right)=0 \tag{27}$$

subject to the initial conditions $f^{(k)}(0)=c_k$, for $k=0,1,2,\cdots,p-1$.

Let [0, T] be the interval over which we want to find the solution of the initial-value problem (27). In actual applications of the DTM, the approximate solution of the initial value problem (27) can be expressed by the following finite series:

$$f(t)=\sum_{n=0}^{N}a_n(t-t_0)^n,\ t\in[0,T] \tag{28}$$

The multi-step approach introduces a new idea for constructing the approximate solution. Assume that the interval [0, T] is divided into M subintervals $[t_{m-1},t_m]$, $m=0,1,2,\cdots,M$ of equal step size h = (T/M) by using the nodes t_m = mh. The main ideas of the MDTM are as follows. First, we apply the DTM to Equation (24) over the interval $[0,t_1]$, we will obtain the following approximate solution:

$$f_1(t)=\sum_{n=0}^{N} a_{1n}(t-t_0)^n,\ t\in[0,t_1] \tag{29}$$

using the initial conditions $f_1^{(k)}(0)=c_k$. For $m\geq 2$ and at each subinterval $[t_{m-1},t_m]$ we will use the initial conditions $f_m^{(k)}(t_{m-1})=f_{m-1}^{(k)}(t_{m-1})$ and apply the DTM to Equation (27) over the interval $[t_{m-1},t_m]$, where t_0 in Equation (23) is replaced by t_{m-1}. The process is repeated and generates a sequence of approximate solutions $f_m(t), m=1,2,\cdots,M$, for the solution $f(t)$:

$$f_m(t)=\sum_{n=0}^{N} a_{mn}(t-t_{m-1})^n,\quad t\in[t_m,t_{m-1}] \tag{30}$$

where $N=K\cdot M$. In fact, the MDTM assumes the following solution:

$$f(t)=\begin{cases} f_1(t), & t\in[0,t_1] \\ f_2(t), & t\in[t_1,t_2] \\ \vdots & \\ f_M(t), & t\in[t_{M-1},t_M] \end{cases} \tag{31}$$

The new algorithm, MDTM, is simple for computational performance for all values of h. It is easily observed that if the step size h = T, then the MDTM reduces to the classical DTM. As we will see in the next section, the main advantage of the new algorithm is that the obtained series solution converges for wide time regions and can approximate non-chaotic or chaotic solutions.

3.3. Analytical Solution Be the MDTM

By applying the MDTM to Equations (18)-(21), gives the following recursive relations in each sub-domain (t_i, t_{i+1}), $i = 0,1,\cdots,N-1$.

$$\left(1+\frac{1}{\beta}\right)(k+1)(k+2)(k+3)F[k+3]+\sum_{r=0}^{k}(k-r+1)(k-r+2)F[r]F[k-r+2]$$
$$+\gamma\left(\delta[k]-\sum_{r=0}^{k}(r+1)(k-r+1)F[r+1]F[k-r+1]\right)+M\left[\sum_{r=0}^{k}(k-r+1)(k-r+2)G[r]G[k-r+2]\right]$$
$$+M\left[\gamma\left(\delta[k]-\sum_{r=0}^{k}(r+1)(k-r+1)G[r+1]G[k-r+1]\right)\right]=0$$

(32)

$$\lambda(k+1)(k+2)(k+3)G[k+3]+\sum_{r=0}^{k}(k-r+1)(k-r+2)F[r]G[k-r+2]$$
$$-\sum_{r=0}^{k}(k-r+1)(k-r+2)G[r]F[k-r+2]=0$$

(33)

$$\left(1+\frac{4}{3}R_a\right)(k+1)(k+2)\theta[k+2]+P_r\sum_{r=0}^{k}(k-r+1)F[r]\theta[k-r+1]$$
$$+P_rE\left(1+\frac{1}{\beta}\right)_r\sum_{r=0}^{k}(r+2)(r+1)(k-r+2)(k-r+1)F[r+2]F[k-r+2]$$
$$+P_rD_u(k+1)(k+2)\varnothing[k+2]=0$$

(34)

$$(k+1)(k+2)\varnothing[k+2]+S_c\sum_{r=0}^{k}(k-r+1)F[r]\varnothing[k-r+1]+S_cS_o(k+1)(k+2)\theta[k+2]=0$$

(35)

where $F[k], G[k], \theta[k]$ and $\varnothing[k]$ are the differential transform of $f(\eta), g(\eta), \theta(\eta)$ and $\varnothing(\eta)$.

The differential transformed boundary conditions in Equation (22) to:

$$F[0]=0,\ G[0]=0,\ \theta[0]=1,\ \varnothing[0]=1,$$
$$F[1]=\lambda,\ F[2]=f_2,\ G(1)=0,\ G[2]=g_2,\ \theta[1]=t_1,\ \varnothing[1]=c_1.$$

(36)

where f_2, g_2, t_1 and c_1 are constants. These constants are computed from the boundary condition. Moreover, substituting Equation (36) into Equation (32)-(35) and by using the recursive method, we can calculate other values of $F(k), G(k), \theta(k)$ and $\varnothing(k)$. Hence, substituting all $F(k), G(k), \theta(k)$ and $\varnothing(k)$, into Equation (26), we obtain series solutions.

The velocity, induced magnetic field, temperature and concentration distributes achieved with the aid of MATHAMATICA application software.

3.4. Numerical Solutions by Using the FDM

To solve the ordinary differential Equations (18)-(21), we use a finite difference based numerical algorithm. We reduce the order of Equations (18) and (19) by one with the help of the substitution:

$$q=\frac{df}{d\eta}=f',\quad p=\frac{dg}{d\eta}=g'$$

(37)

Equations (18)-(21) in view of Equation (37) can be written as:

$$\left(1+\frac{1}{\beta}\right)q''+fq'+\gamma\left(1-q^2\right)-M\left[gp'+\gamma\left(1-p'^2\right)\right]=0 \tag{38}$$

$$\lambda p''+fp'-pq'=0 \tag{39}$$

$$\left(1+\frac{4}{3}R_a\right)\theta''+P_rf\theta'+P_rE\left(1+\frac{1}{\beta}\right)q'^2+P_rD_u\varnothing''=0 \tag{40}$$

$$\varnothing''+S_cf\varnothing'+S_cS_o\theta''=0 \tag{41}$$

The boundary conditions:

$$\begin{gathered}f(0)=0,\ q(0)=\delta,\ g(0)=0,\ p(0)=0,\ \theta(0)=1,\ \varnothing(0)=1,\\ q(\infty)=1,\ p(\infty)=1,\ \theta(0)=0,\ \varnothing(\infty)=0.\end{gathered} \tag{42}$$

Now, we can write Equations (37)-(41) in the following finite difference form

$$\left(1+\frac{1}{\beta}\right)\left(\frac{q_{i+1}-2q_i+q_{i-1}}{(\Delta\eta)^2}\right)+f_i\left(\frac{q_{i+1}-q_{i-1}}{2\Delta\eta}\right)+\gamma\left(1-q_i^2\right)-M\left[g_i\left(\frac{p_{i+1}-p_{i-1}}{2\Delta\eta}\right)+\gamma\left(1-p_i^2\right)\right]=0 \tag{43}$$

$$\lambda\left(\frac{p_{i+1}-2p_i+p_{i-1}}{(\Delta\eta)^2}\right)+f_i\left(\frac{p_{i+1}-p_{i-1}}{2\Delta\eta}\right)-g_i\left(\frac{q_{i+1}-q_{i-1}}{2\Delta\eta}\right)=0 \tag{44}$$

$$\begin{gathered}\left(1+\frac{4}{3}R_a\right)\left(\frac{\theta_{i+1}-2\theta_i+\theta_{i-1}}{(\Delta\eta)^2}\right)+P_rf_i\left(\frac{\theta_{i+1}-\theta_{i-1}}{2\Delta\eta}\right)+P_rE\left(1+\frac{1}{\beta}\right)\left(\frac{q_{i+1}-q_{i-1}}{2\Delta\eta}\right)^2\\ +P_rD_u\left(\frac{\varnothing_{i+1}-2\varnothing_i+\varnothing_{i-1}}{(\Delta\eta)^2}\right)=0\end{gathered} \tag{45}$$

$$\left(\frac{\varnothing_{i+1}-2\varnothing_i+\varnothing_{i-1}}{(\Delta\eta)^2}\right)+S_c f_i\left(\frac{\varnothing_{i+1}-\varnothing_{i-1}}{2\Delta\eta}\right)+S_c S_o\left(\frac{\theta_{i+1}-2\theta_i+\theta_{i-1}}{(\Delta\eta)^2}\right)=0$$

(46)

The velocity, induced magnetic field, temperature and concentration distributes at all interior nodal points computed by successive applications of the above finite difference equations and these are achieved with the aid of MATLAB application software.

4. RESULTS AND DISCUSSION

The system of the equations which describe the motion of Casson fluid in the boundary layer with heat and mass transfer and in the presence of induced magnetic field with radiation are solved analytically by using (DTM) and numerically by using (FDM). The values of the velocity, temperature, concentration and induced magnetic field are obtained as a functions of the physical parameters of the problem such as Casson parameter β, wedge parameter γ, magnetic parameter M, magnetic Prandtl number λ, radiation parameter R_a, Prandtl number P_r, Eckert number E, Dufour parameter D_u, Schmidt number S_c, and Soret number S_0. These effects are illustrated graphically through a set of (1 - 18).

Figures 1-5 represent the velocity profiles for the flow parameters. The effect of β on the velocity profiles is presented in Figure 1. It is seen that the velocity increases when β increases. Figure 2clears the effect of γ on the velocity profiles. It is shown that the velocity increases with increasing γ. Figure 3 clears the effect of M on the velocity profiles. It is shown that the velocity decreases with increasing M, also it decrease with increasing λ as shown in Figure 4. The effect of δ on the velocity profiles is presented in Figure 5. It is shown that the velocity increases with increasing δ. Figures 6-10 represent the induced magnetic field profiles for the flow parameters. It is observed that the induced magnetic field increases with increasing β, γ and δ, while decreases with increasing M and λ. Figures 11-15 represent the temperature profiles for the flow parameters. It is observed that the temperature increases with increasing R_a, E and D_u, while decreases with increasing β and P_r. Figures 16-18 represent the concentration profiles for the flow parameters. It is noted that concentration increases with an increases S0, while it decrease with the increases β and Sc.

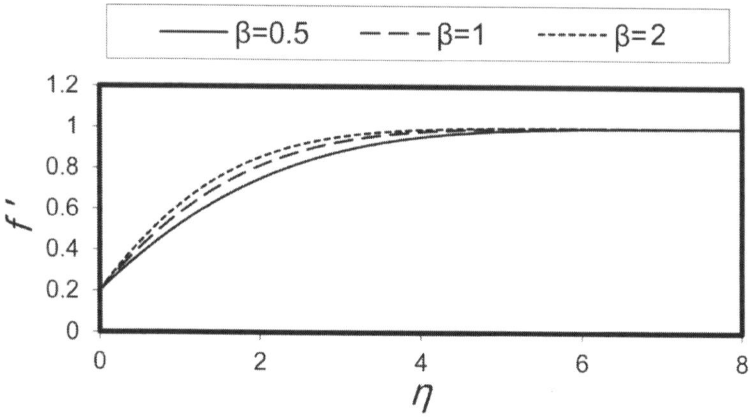

Figure 1. Velocity profiles $f'(\eta)$ for some values of β.

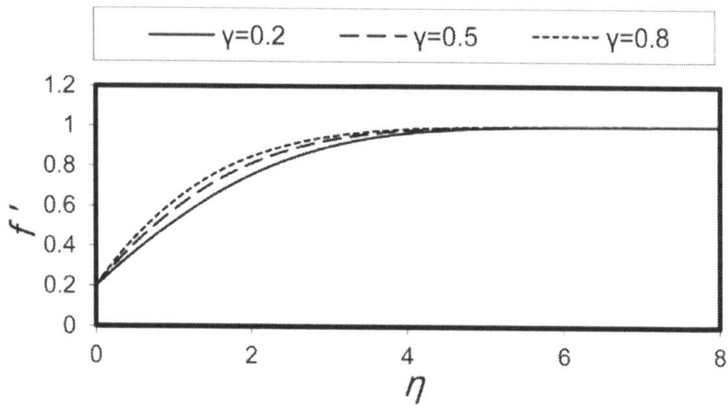

Figure 2. Velocity profiles $f'(\eta)$ for some values of γ.

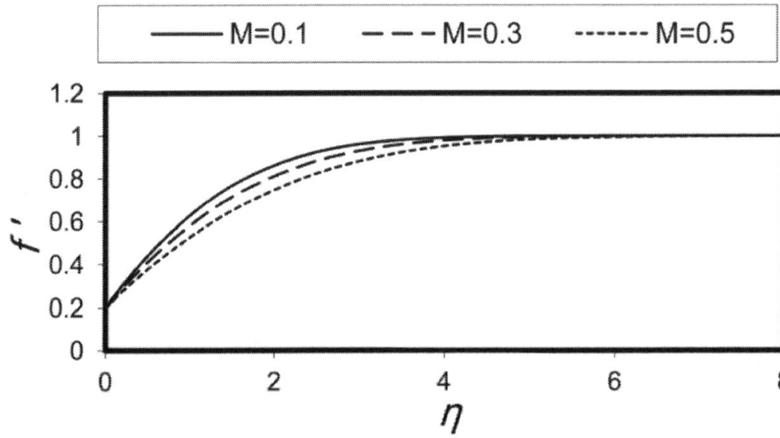

Figure 3. Velocity profiles $f'(\eta)$ for some values of M.

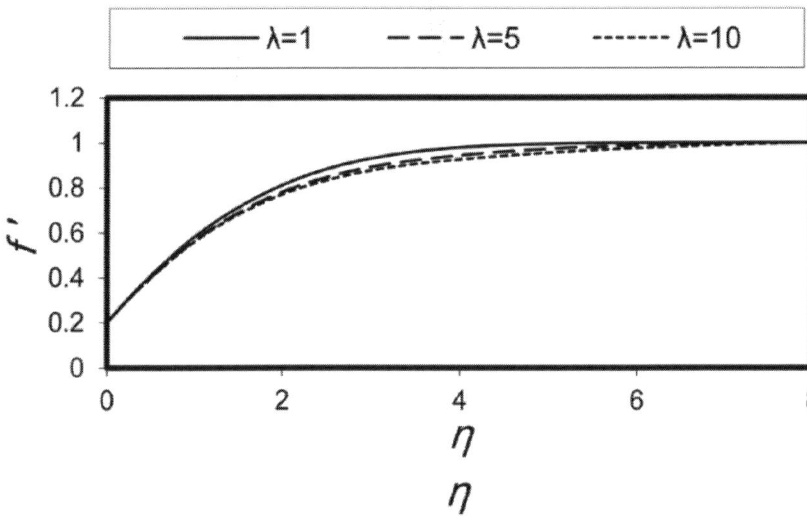

Figure 4. Velocity profiles $f'(\eta)$ for some values of λ.

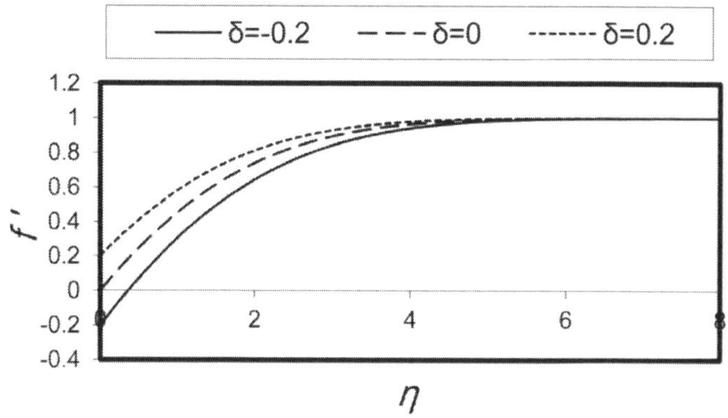

Figure 5. Velocity profiles $f'(\eta)$ for some values of δ.

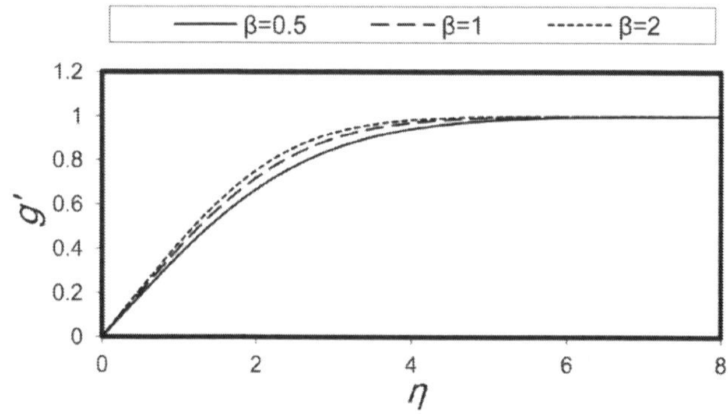

Figure 6. Induced magnetic profiles $g'(\eta)$ for some values of β.

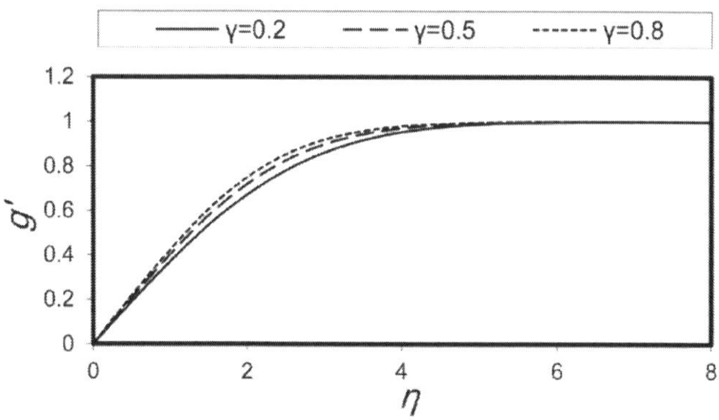

Figure 7. Induced magnetic profiles $g'(\eta)$ for some values of γ.

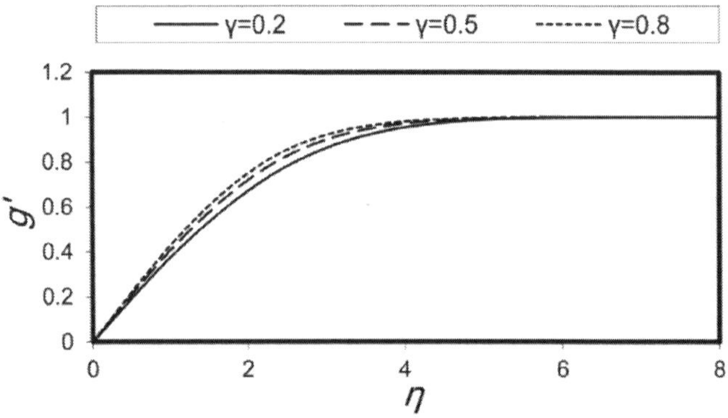

Figure 8. Induced magnetic profiles $g'(\eta)$ for some values of M.

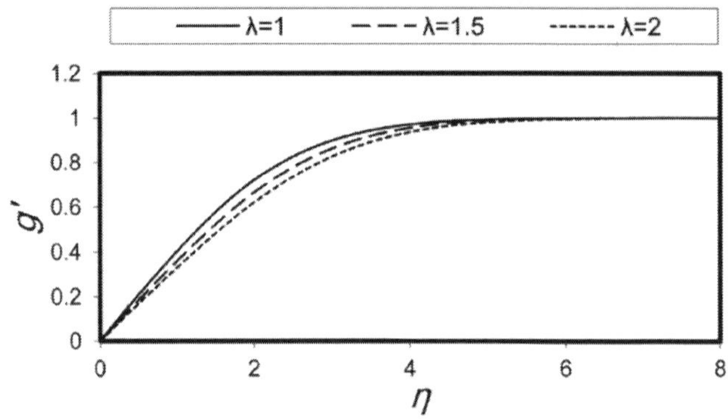

Figure 9. Induced magnetic profiles $g'(\eta)$ for some values of λ .

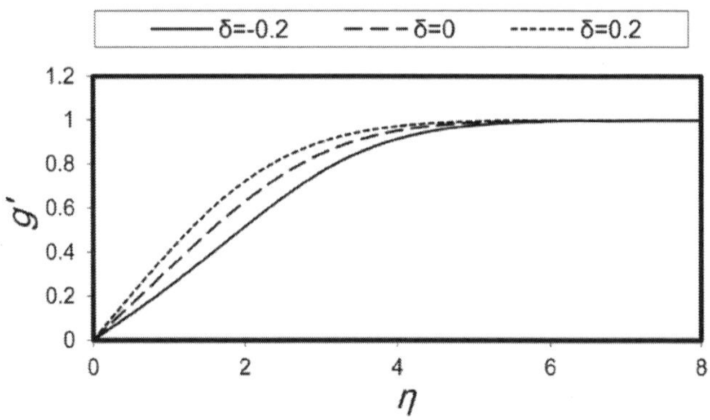

Figure 10. Induced magnetic profiles $g'(\eta)$ for some values of δ .

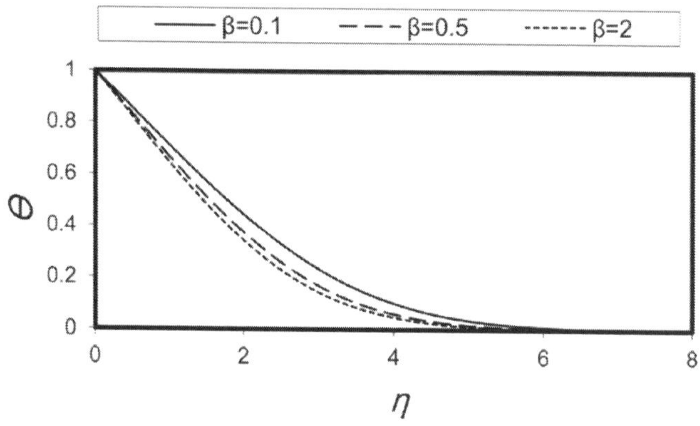

Figure 11. Temperature profiles $\theta(\eta)$ for some values of β .

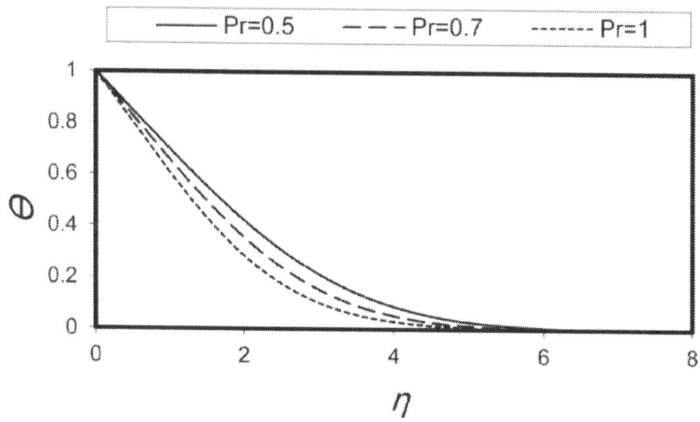

Figure 12. Temperature profiles $\theta(\eta)$ for some values of P_r.

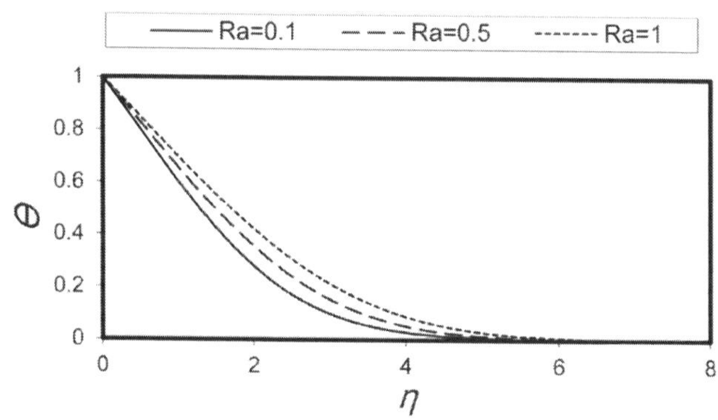

Figure 13. Temperature profiles $\theta(\eta)$ for some values of R_a.

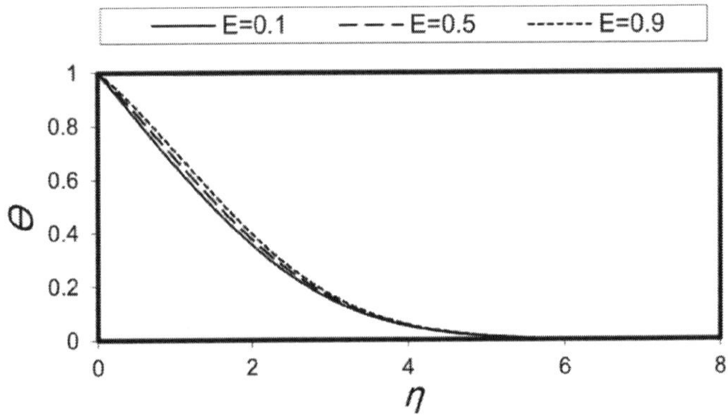

Figure 14. Temperature profiles $\theta(\eta)$ for some values of E.

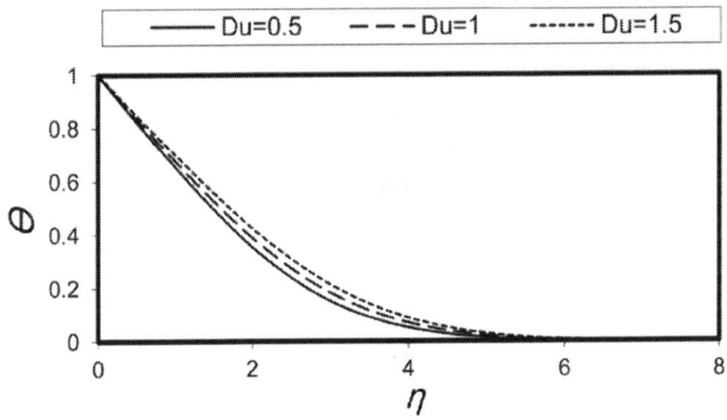

Figure 15. Temperature profiles $\varnothing(\eta)$ for some values of D_u.

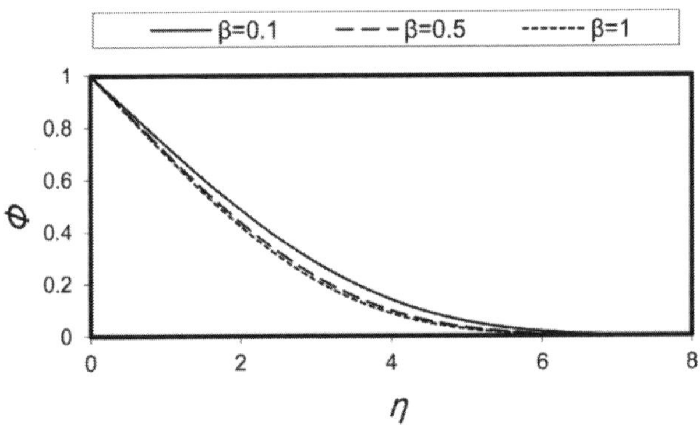

Figure 16. Concentration profiles $\varnothing(\eta)$ for some values of β.

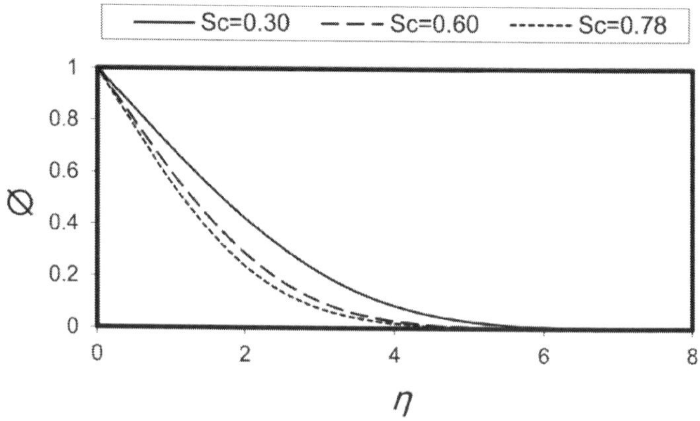

Figure 17. Concentration profiles $\varnothing(\eta)$ for some values of S_c.

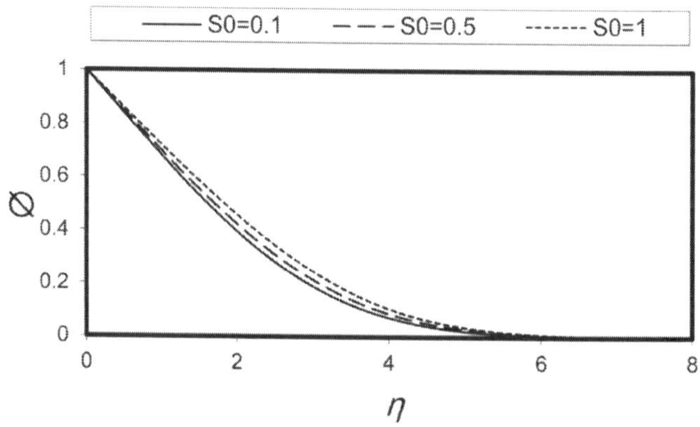

Figure 18. Concentration profiles $\varnothing(\eta)$ for some values of S_0.

In order to verify the accuracy of the numerical solution of present work by using (FDM) we have compared these results for $f''(0)$ when $\delta = 0$ and M = 0 with Rajagopal et al. [16] , Kuo [17] , Ishak et al. [18] , K. Jafar [8] . It is observed that this numerical solution in excellent agreement with the results previous obtained values as shown in Table 1. Finally, Tables 2-4 shows comparison between numerical solution by using (FDM) and the analytical solution using (DTM). It is observed that this approximate numerical solution is in excellent agreement with the results of the analytical solution of by using (DTM).

Table 1. Comparison between the previous obtained values of $f''(0)$ by Rajagopal et al. [16] , Kuo [17] , Ishak et al. [18] , K. Jafar [8] and our results when δ =0 and M =0.

γ	Rajagopal *et al.* [16]	Kuo [17]	Ishak *et al.* [18]	K. Jafar [8]	Present result
0.0	-	0.469600	0.4696	0.4696	0.46960075
0.1	0.587035	0.587880	0.5870	0.5871	0.5870358
0.3	0.774755	0.775524	0.7748	0.7748	0.7747554
0.5	0.927680	0.927905	0.9277	0.9277	0.9276813
1.0	1.232585	1.231289	1.2326	1.2326	1.232590

Table 2. Comparison between the previous obtained values of $f(n)$ and $f'(n)$ where $\beta=1$, $\gamma=0.5$, $M=0.3$, $\lambda=1$, $\delta=0.2$, $P_r=0.7$, $R_a=0.5$, $E=0.1$, $D_u=0.5$, $S_c=0.3$ and $S_o=0.5$

η	$f(\eta)$			$f'(\eta)$		
	DTM	MDTM	FDM	DTM	MDTM	FDM
0.0	0.0	0.0	0.0	0.2	0.2	0.2
1.0	0.403766	0.403766	0.4037649	0.580616	0.580616	0.58061605
2.0	1.11098	1.11098	1.1109816	0.811421	0.811421	0.81142107
3.0	1.98901	1.98901	1.9890103	0.929608	0.929609	0.92960949
4.0	2.95048	2.94735	2.9473479	1.0034	0.97895	0.97895070
5.0	6.32369	3.93601	3.9360082	14.7103	0.995059	0.99505878
6.0	415.231	4.93359	4.9335872	1902.95	0.999104	0.99910402
7.0	28673.5	5.93319	5.9331922	111857	0.999884	0.99988388
8.0	1.0757×10^6	6.93315	6.9331527	3.62989×10^6	1.0	1.0

Table 3. Comparison between the previous obtained values of $g(\eta)$ and $g'(\eta)$ where β=1, γ= 0.5,M=0.3, λ=1, δ=0.2,P_r = 0.7,R_a=0.5,*E=0.1, D_u=0.5*,S_c=0.3 and S_o=0.5.

η	$g(\eta)$			$g'(\eta)$		
	DTM	MDTM	FDM	DTM	MDTM	FDM
0.0	0.0	0.0	0.0	0.0	0.0	0.0
1.0	0.20504	0.20504	02.050395	0.402848	0.402848	0.4028479
2.0	0.776432	0.776432	0.7764307	0.718855	0.718855	0.7188556
3.0	1.59576	1.59579	1.5957888	0.897193	0.897483	0.974845
4.0	2.45379	2.53591	2.5359114	0.411927	0.970245	0.9702454
5.0	−31.0684	3.51998	3.5199796	−185.501	0.993106	0.9931059
6.0	−4620.89	4.51661	4.5166091	−20617.8	0.998756	0.9987563
7.0	−281829	5.51606	5.5160617	-1.07009×10^6	0.99984	0.9998396
8.0	-9.71748×10^6	6.51601	6.5160072	-3.21308×10^7	1.0	1.0

Table 4. Comparison between the previous obtained values of $\theta(\eta)$ and $\varnothing(\eta)$ where β=1, γ= 0.5,M=0.3, λ=1, δ=0.2,P_r = 0.7,R_a=0.5,E=0.1, D_u=0.5,S_c=0.3 and S_o=0.5.

η	$\theta(\eta)$			$\varnothing(\eta)$		
	DTM	MDTM	FDM	DTM	MDTM	FDM
0.0	1.0	1.0	1.0	1.0	1.0	1.0
1.0	0.65649	0.65649	0.6564895	0.696176	0.696176	0.6961763
2.0	0.355842	0.355842	0.3558407	0.422562	0.422562	0.4225611
3.0	0.153485	0.15349	0.1534889	0.213823	0.213822	0.2138215
4.0	0.0417624	0.052241	0.052240	0.0885228	0.0866998	0.0866989
5.0	−3.56762	0.0141772	0.01417667	0.649388	0.0273532	0.0273525
6.0	−393.027	0.00306269	0.00306253	67.9412	0.00653717	0.00653686
7.0	−20071.9	0.000478198	0.000478176	3451.29	0.00107547	0.00107538
8.0	−592012	0.0	0.0	101282	0.0	0.0

5. CONCLUSIONS

In this work we have obtained the numerical solution of MHD boundary layer flow of non-Newtonian Casson fluid on a moving wedge with heat and mass transfer and induced magnetic field. The effects of thermal diffusion and diffusion thermo with induced magnetic field are taken into consideration. The resulting partial differential equations which describe the problem are transformed into ordinary differential equations by using a similarity transformation and then solved numerically by using the finite difference method (FDM) and analytically by using (DTM). A representative set of numerical results for velocity, induced magnetic field, temperature and concentration profiles is presented graphically and discussed. The figures and tables clearly show that the results by using (FDM) are in excellent agreement with previously published works and with the results of the analytical solution by using (DTM). The important results for this study summarized as follows:

- The velocity distribution decreases with the increase of M and λ, while it increases with an increase of β, γ and δ.
- The temperature distribution increases with the increase of R_a, E and D_u, while it decreases with the increases of P_r and β.
- The induced magnetic field distribution increases with the increase of M and λ, while it decreases with the increases of β.

Ÿ The concentration distribution increases with increase of S_0, while it decreases with the increases of S_c

REFERENCES

1. Raptis, A. and Perdikis, C. (1984) Free Convection under the Influence of a Magnetic Field. Nonlinear Analysis: Theory, Methods & Applications, 8, 749-756.

2. Kumari, M., Takhar, H.S. and Nath, G. (1990) MHD Flow and Heat Transfer over a Stretching Surface with Prescribed Wall Temperature or Heat Flux. Wärme - und Stoffübertragung, 25, 331-336.

3. Takhar, H.S., Kumari, M. and Nath, G.G. (1993) Unsteady Free Convection Flow under the Influence of a Magnetic Field. Archive of Applied Mechanics, 63, 313-321.

4. Ali, F.M. Nazar, R., Arifin, N.M. and Pop, I. (2011) MHD Stagnation-Point Flow and Heat Transfer towards Stretching Sheet with Induced Magnetic Field. Applied Mathematics and Mechanics, 32, 409-418.

5. Ali, K., Ashraf, M., Ahmad, S. and Batool, K. (2012) Viscous Dissipation and Radiation Effects in MHD Stagnation Point Flow towards a Stretching Sheet with Induced Magnetic Field. World Applied Sciences Journal, 16, 1638-1648.

6. Ali, K. and Ashraf, M. (2012) Thermal Reversal in MHD Stagnation Point Flow towards a Stretching Sheet with Induced Magnetic Field and Viscous Dissipation Effects. World Applied Sciences Journal, 16, 1615-1625.

7. Ali, F.M., Nazar, R., Arifin, N.M. and Pop, I. (2011) MHD mixed Convection Boundary Layer Flow toward a Stagnation Point on a Vertical Surface with Induced Magnetic Field Was Studied. Journal of Heat Transfer, 133, Article ID: 022502, 1-6.

8. Jafar, K., Nazar, R., Ishak, A. and Pop, I. (2013) MHD Boundary Layer Flow Due to a Moving Wedge in a Parallel Stream with the Induced Magnetic Field. J. Boundary Value Problems, 1-14.

9. Nakayama, M. and Sawada, T. (1988) Numerical Study on the Flow of a Non-Newtonian Fluid through an Axisymmetric Stenosis. Journal of Biomechanical Engineering, 110, 137-143.

10. Abdelnaby, M.A., Eldabe, N.T. and Abou-zeid, M.Y. (2006) Numerical Study of Pulsatile MHD Non-Newtonian Fluid Flow With heat and Mass Transfer through a Porous Medium between Two Permeable Parallel Plates. Ind. J. Mech. Cont. & Math Sci, 1, 1-15.

11. Eldabe, N.T., El-Sakka, A.G. and Fouda, A. (2002) Numerical Treatment of the MHD Convective Heat and Mass Transfer in an Electrically Conducting Fluid over an Infinite Solid Surface in Presence of Internal Heat Generation. Zeitschrift für Naturforschung, 58, 601-611.

12. Eldabe, N.T., El-Sayed, M.F., Ghaly, A.Y. and Sayed, H.M. (2007) Peristaltically Induced Transport of a MHD Biviscosity Fluid in a Non-Uniform Tube. Physica A: Statistical Mechanics and its Applications, 383, 253-266.

13. Eldabe, N.T., El-Sayed, M.F., Ghaly, A.Y. and Sayed, H.M. (2008) Mixed Convective Heat and Mass Transfer in a Non-Newtonian Fluid at a Peristaltic Surface with Temperature-Dependent Viscosity. Archive of Applied Mechanics, 78, 599-624.

14. Eldabe, N.T. and Hassan, A.A. (1991) Non-Newtonian-Flow Formation in Couette Motion in Magnethohydrodynamics with Time-Varying Suction. Canadian Journal of Physics, 69, 75-85.

15. Zhou, J.K. (1986) Differential Transformation and Its Application for Electrical Circuits. Huazhong University Press, Wuhan.

16. Rajagopal, K.R., Gupta, A.S. and Nath, T.Y. (1983) A Note on the Falkner-Skan Flows of a Non-Newtonian Fluid. International Journal of Non-Linear Mechanics, 18, 313-320.

17. Kuo, B.L. (2003) Application of the Differential Transformation Method to the Solutions of Falkner-Skan Wedge Flow. Acta Mechanica, 164, 161-174.

18. Ishak, A., Nazar, R. and Pop, I. (2009) MHD Boundary Layer Flow past a Moving Wedge. Magnetohydrodynamics, 45, 3-10.

Chapter 8

Perturbation analysis of magnetohydrodynamics oscillatory flow on convective-radiative heat and mass transfer of micropolar fluid in a porous medium with chemical reaction

Dulal Pal[a], Sukanta Biswas[b]

[a] Department of Mathematics, Visva-Bharati University, Siksha-Bhavana, Santiniketan, West Bengal 731 235, India

[b] Bhedia High School, Bhedia, Burdwan, West Bengal 713 126, India

ABSTRACT

This paper deals with the perturbation analysis of mixed convection heat and mass transfer of an oscillatory viscous electrically conducting micropolar fluid over an infinite moving permeable plate embedded in a saturated porous medium in the presence of transverse magnetic field.

Analytical solutions are obtained for the governing basic equations. The effects of permeability, chemical reaction, viscous dissipation, magnetic field parameter and thermal radiation on the velocity distribution, micro-rotation, skin friction and wall couple stress coefficients are analyzed in detail. The results indicate that the effect of increasing the chemical reaction has a tendency to decrease the skin friction coefficient at the wall, while opposite trend is seen by increasing the permeability parameter of the porous medium. Also micro-rotational velocity distribution increases with an increase in the magnetic field parameter.

KEYWORDS

Chemical reaction; Heat and mass transfer; Micropolar fluid; Magneto-hydrodynamics; Porous medium; Thermal radiation; Viscous dissipation

1. INTRODUCTION

There has been a considerable amount of interest among researchers to study the convective flow with simultaneous heat and mass transfer under the influence of a magnetic field and chemical reaction as such processes exist in many branches of science and technology. In many industries application of such type of problems are seen in the chemical industry, cooling of nuclear reactors and magnetohydrodynamic (MHD) power generators. The study of MHD mixed convection heat and mass transfer with chemical reaction are of great importance to engineers and scientists because of its almost universal occurrence in many branches of science and engineering, and hence received a considerable amount of attention in recent years. The study of heat transfer in porous medium in the presence of chemical reaction has important engineering applications e.g. oxidation of solid materials, synthesis of ceramic materials and tubular reactors. There are two types of reactions such as (i) homogeneous reaction and (ii) heterogeneous reaction. A homogeneous reaction occurs uniformly throughout the given phase, whereas heterogeneous reaction takes place in a restricted region or within the boundary of a phase. The effects of a chemical reaction depend greatly on whether the reaction is heterogeneous or homogeneous. A chemical reaction is said to be first-order if the rate of reaction is directly proportional to the concentration itself. In many industrial processes involving flow and mass transfer over a moving surface, the diffusing species can be generated or absorbed due to some kind of chemical reaction with the ambient fluid which can greatly affect the flow and hence the properties and quality of the final product. These processes take place in numerous industrial applications, such as in the polymer production and in manufacturing of ceramics or glassware, and food processing chemical reaction that occurs between a foreign mass and a fluid in which a plate is moving. Ibrahim et al. [1] studied the effect of chemical reaction and thermal radiation absorption on the unsteady MHD free convection flow past a semi-infinite vertical permeable moving plate with heat source and suction. They found that the velocity profiles and concentration profile increased due to a decrease in the chemical reaction parameter. Al-Odat and Al-Azab [2] studied the influence of chemical reaction on transient MHD free convection over a moving vertical plate. They found that the velocity as well as concentration decrease with increasing the chemical reaction parameter. Seddeek et al. [3] examined the effects of chemical reaction and variable viscosity on hydromagnetic mixed convection heat and mass transfer for Hiemenz flow through porous media with radiation, and they found that the local Sherwood number significantly increases with the chemical reaction parameter. Pal and Talukdar [4] used perturbation analysis to study unsteady magnetohydrodynamic convective heat and mass transfer in a boundary layer slip flow past a vertical permeable plate with thermal radiation and chemical reaction. They solved the non-linear coupled partial differential equations by perturbation technique and found that the velocity as well as concentration decrease with increasing the chemical reaction parameter. Hsiao [5] studied the heat and mass transfer on

MHD mixed convection of viscoelastic fluid past a stretching sheet with Ohmic dissipation. Rout et al. [6] studied the effect of thermal radiation and chemical reaction on double diffusive natural convective MHD flow through a porous medium. They considered destructive reaction in this paper, i.e., when chemical reaction parameter is positive then the concentration decreases due to the contribution of mass diffusion in concentration equation, while in the generative reaction, i.e., when chemical reaction parameter is negative, they observed the reverse effect. They further found that with increasing chemical reaction parameter there is substantial increase in the temperature profile.

Mixed convection in porous media has gained significant attention for its importance in engineering applications such as in the design of nuclear reactor, ceramics processing, geothermal systems, solid matrix heat exchangers, thermal insulations, crude oil drilling and compact heat exchanges etc. Convection in porous media can also be applied to underground coal gasification, ground water hydrology, iron blast furnaces, wall cooled catalytic reactors, energy efficient drying processes, cooling of nuclear fuel in shipping flasks, cooling of electronic equipments and natural convection in the Earth's crust. The fundamental problem of flow through and past porous media has been studied extensively over the past few years both theoretically and experimentally. Modather et al.[7] analytically studied MHD heat and mass transfer oscillatory flow of a micropolar fluid over a vertical permeable plate in a porous medium, and the results indicate that increasing the permeability parameter produces an increasing effect on the skin friction coefficient and the couple stress coefficient at the wall. El-Hakiem [8] examined the MHD oscillatory flow on free convection-radiation through a porous medium with constant suction velocity. They observed that the velocity increases when the permeability of the porous medium is increased. Pal and Talukdar [9] investigated the buoyancy and chemical reaction effects on MHD mixed convection heat and mass transfer in a porous medium with thermal radiation and Ohmic heating, and they concluded that the presence of porous medium increases the skin friction coefficient, whereas the effect of increasing the value of porous permeability decrease the value of the local Nusselt number. Sugunamma et al. [10] studied the inclined magnetic field and chemical reaction effects on flow over a semi-infinite vertical porous plate through porous medium. They solved the non-linear and coupled governing equations by adopting a perturbative series expansion about a small parameter, ε, and they observed that the velocity gradient at the surface increases with a decrease in the porosity parameter. Acharya et al. [11] analyzed the free convective fluctuating MHD flow through porous media past a vertical porous plate with variable temperature and heat source, and they recorded that the presence of porous media has no significant contribution to the flow characteristics whereas viscous dissipation compensates for the heating and cooling of the plate due to convective current.

The study of free convection flow with magnetic field plays a major role in liquid metals, electrolytes and ionized gases and thermal physics of hydromagnetic problems with mass transfer have enormous applications in power engineering. Prasad et al. [12]examined the influence of internal heat generation/absorption, thermal radiation, magnetic field, variable fluid property and viscous dissipation on heat transfer characteristics of a Maxwell fluid over a stretching sheet, and they have pointed out that the horizontal velocity decreases with an increase in the magnetic field parameter. They concluded that this is due to the fact that the transverse magnetic field has a tendency to create a drag like force, known as the Lorentz force to resist the flow. Vija et al. [13]studied the effects of induced magnetic field and viscous dissipation on MHD mixed convective flow past a vertical plate in the presence of thermal radiation. They found that the values of induced magnetic field remained negative, i.e. induced magnetic flux reversal arises for all distances in the boundary layer.

The effects of radiation on MHD flow and heat transfer problems have become industrially more important. The thermal radiation effects become intensified at high absolute temperature levels due to basic difference between radiation and the convection and conduction energy-exchange mechanisms. Many engineering processes occur at high temperatures and hence the knowledge of thermal radiation heat transfer is essential for designing appropriate equipments such as nuclear power plants, gas turbines and various propulsion devices for aircraft, missiles and satellites. When radiative heat transfer takes place in the electrically conducting fluid, it is ionized due to the high operating temperature. In view of these, many authors have made contributions to the study of fluid flow with thermal radiation. Hsiao [14] analyzed heat and mass transfer of micropolar fluid flow in the presence of thermal radiation past a nonlinearly stretching sheet. Shateyi et al. [15] investigated the effects of thermal radiation, Hall currents, Soret and Dufour number on MHD mixed convection flow over a vertical surface in porous media, and they found that the fluid temperature increases due to an increase in the thermal radiation. Also they found that the concentration decreases as the radiation parameter value is increased. Pal and Mondal [16] examined the effects of Soret Dufour, chemical reaction and thermal radiation on MHD non-Darcy unsteady mixed convective heat and mass transfer over a stretching sheet, and they concluded that the effect of thermal radiation is to increase temperature in the thermal boundary layer. The effects of thermal radiation and viscous dissipation on MHD heat and mass diffusion flow past an oscillating vertical plate embedded in a porous medium with variable surface conditions were studied by Kishore et al. [17]. Thermal radiation effects on MHD convective flow over a plate in a porous medium was studied by Karthikeyan et al. [18] by using perturbation technique, and it was observed that the increase in the radiation parameter implies the decrease in the boundary layer thickness and enhances the rate of heat transfer. Hossain and Samand [19] examined the heat and mass transfer of a MHD free convection flow along a stretching sheet with chemical reaction, thermal radiation and heat generation in the presence of magnetic field. They conclude that the concentration profiles increase as the values of the radiation parameter is increased. Recently, Hsiao [20] performed an analysis to study the combined effects mixed convection and thermal radiation in nanofluid with multimedia physical features.

Viscous dissipation plays an important role in changing the temperature distribution, just like an energy source, which affects the heat transfer rates considerably. In fact, the shear stresses can induce a considerable amount of heat generation. El-Aziz [21]investigated the mixed convection flow of a micropolar fluid from an unsteady stretching surface with viscous dissipation. He concluded that the viscous dissipation produces heat due to drag between the fluid particles and this extra heat causes an increase of the initial fluid temperature. This increase in the temperature causes an increase in the buoyant force. Kishore et al. [22] studied the influence of chemical reaction and viscous dissipation on unsteady MHD free convection flow past an exponentially accelerated vertical plate with variable surface conditions. They examined that the increase in the viscous dissipation enhanced the fluid temperature, also the rate of heat transfer fell with increasing the Eckert number. Singh and Singh [23] analyzed MHD flow with viscous dissipation and chemical reaction over a stretching porous plate in porous medium. They drew a conclusion that the effect of viscous dissipation on the temperature distribution is insignificant.

The theory of micropolar fluids was originally developed by Eringen [24] which has become a popular field of research in recent times. The concept of micropolar fluid deals with a class of fluids that exhibits certain microscopic effects arising from the micromotions of the fluid elements. These fluids contain dilute suspension of rigid macromolecules with individual motions that support stress and body moments are influenced by spin inertia. Micropolar fluids are those which contain micro-constituents that can undergo rotation, which can affect the hydrodynamics of the flow so that it can be distinctly non-Newtonian. Thus, micropolar fluids are those consisting of randomly oriented particles suspended in a viscous fluid, which can undergo a rotation. Haque et al. [25] analyzed the micropolar fluid behaviors on steady MHD free convection and mass transfer flow with constant heat and mass fluxes, Joule heating and viscous dissipation. It was concluded by them that the motion of micropolar fluid is more for lighter particles and air than heavier particles and water, respectively and the angular motion of microlpolar fluid is greater for heavier particles and water than lighter particles and air, respectively also the micropolar fluid temperature is more for air than water. Sudheer Babu et al. [26]studied the mass transfer effects on unsteady MHD convection flow of micropolar fluid past a vertical moving porous plate through a porous medium with viscous dissipation. Gupta et al. [27] investigated unsteady mixed convection flow of micropolar fluid over a porous shrinking sheet using finite element method.

Different authors studied mass transfer with or without radiation and viscous dissipation effects on the flow past oscillating vertical plate by considering different surface conditions, but the study on the effects of magnetic field on free convection heat and mass transfer with thermal radiation, viscous dissipation, chemical reaction and variable surface conditions in flow through an oscillating plate has not been found in literature and hence the motivation to undertake the present study. Thus the present investigation is concerned with the study of combined effects of magnetic field and first-order chemical reaction in two-dimensional MHD flow, heat and mass transfer of a viscous incompressible fluid past a permeable vertical plate embedded in a porous medium in the presence of viscous dissipation and thermal radiation using perturbation technique. The effects of various physical parameters on the velocity, temperature and concentration profiles as well as on local skin friction co-efficient, Sherwood number and local Nusselt number are shown graphically. Validation of the analysis has been performed by comparing the present results with those of Modather et al. [7].

2. FORMULATION OF THE PROBLEM

We consider mixed convection and diffusion mass transfer of a viscous incompressible electrically conducting micropolar fluid over an infinite vertical porous moving permeable plate embedded in a porous medium in the presence of viscous dissipation, thermal radiation, heat source/sink and chemical reaction. In Cartesian coordinate system, x-axis is measured along the plate and y-axis normal to the plate. A constant magnetic field is applied in y-direction of strength B_0. It is assumed that the magnetic field is of small intensity so that the induced magnetic field is negligible in comparison to the applied magnetic field. The Joule heating is negligible as the term due to electric dissipation is neglected in the energy equation. The fluid is considered to be gray absorbing emitting or radiating but not scatting medium. The Rosseland approximation is used to describe the radiative heat flux which is negligible in the x-direction in comparison to y -direction. The plate moves continuously with uniform velocity u_p^* in its own plane. It is assumed that the temperature of the surface is held uniform at T_w while the ambient temperature takes the constants value T_∞ so that $T_w > T_\infty$. The species concentration at the surface is maintained uniformly at C_w and that of the ambient fluid is taken as C_∞. First-order chemical reaction is considered in this paper since the rate of reaction is directly proportional to the concentration difference which is associated with the concentration of the species C in the solutal boundary layer and the ambient fluid concentration C_∞[7] and [13]. The chemical reaction is assumed to be irreversible. Under these assumptions, the boundary layer equations of motion, energy and mass-diffusion under the influence of uniform transverse magnetic field and Ohmic dissipation in the presence of heat source or sink, viscous dissipation and thermal radiation are as follows:

Equation (1)

$$\frac{\partial v^*}{\partial y^*} = 0 \tag{1}$$

$$\frac{\partial u^*}{\partial t^*} + v^* \frac{\partial u^*}{\partial y^*} = (v + v_r)\frac{\partial^2 u^*}{\partial y^{*2}} + 2v_r \frac{\partial \omega^*}{\partial y^*} + g\beta_T(T - T_\infty) + g\beta_C(C - C_\infty) - \frac{\sigma B_0^2}{\rho}u^* - \frac{(v + v_r)}{K}u^*, \tag{2}$$

$$\rho j^*\left(\frac{\partial \omega^*}{\partial t^*}+v^*\frac{\partial \omega}{\partial y^*}\right)=\gamma\frac{\partial^2\omega^*}{\partial y^{*2}}, \tag{3}$$

$$\frac{\partial T}{\partial t^*}+v^*\frac{\partial T}{\partial y^*}=\alpha\frac{\partial^2 T}{\partial y^{*2}}-\frac{1}{\rho c_p}\frac{\partial q_r{}^*}{\partial y^*}+\frac{\mu}{\rho c_p}\left(\frac{\partial u^*}{\partial y^*}\right)^2, \tag{4}$$

$$\frac{\partial C}{\partial t^*}+v^*\frac{\partial C}{\partial y^*}=D\frac{\partial^2 C}{\partial y^{*2}}+\gamma_1{}^*(C-C_\infty), \tag{5}$$

where (u^*, v^*) are the component of the velocity at any point (x^*, y^*); ω^* is the component of the angular velocity normal to the x^*y^* plane; T is the temperature of the fluid; and C is the mass concentration of the species in the flow ρ, ν, ν_r, g, β_T, β_C, σ, K, j^*, γ, α, D, and γ_1^* are the density, kinematic viscosity, kinematic rotational viscosity, acceleration due to gravity, coefficient of volumetric thermal expansion of the fluid, coefficient of volumetric mass expansion of the fluid, electrical conductivity of the fluid, permeability of the medium, microinertia per unit mass, spin gradient viscosity, thermal diffusivity, molecular diffusivity, and the dimensional chemical reaction parameter, respectively. It is important to note that the change in the concentration of species gives rise to the solutal buoyancy so there is direct connection between Eq. (2) and Eq. (5) hence these equations are coupled.

The appropriate boundary conditions for the problem are:

$$u^*=u_p{}^*,\quad \omega^*=-n_1\frac{\partial u^*}{\partial y^*},\quad T=T_\infty+\varepsilon(T_\omega-T_\infty)e^{n^*t^*}, \tag{6}$$

$$C=C_\infty+\varepsilon(C_\omega-C_\infty)e^{n^*t^*} \quad \text{at } y^*=0,$$

$$u^*\to 0,\quad \omega^*\to 0,\quad T\to T_\infty,\quad C\to C_\infty \quad \text{as } y^*\to\infty.$$

The following comment should be made about the boundary condition used for the microrotation term: when $n_1 = 0$, we obtain from the boundary condition stated in Eq. (6), for the microrotation, $\omega^* = 0$. This represents the case of concentrated particle flows in which the microelements close to the wall are not able to rotate. The case corresponding to $n_1 = 0.5$ results in the vanishing of the antisymmetric part of the stress tensor and represents weak concentrations. When $n_1 = 1$ then this is the case of turbulent boundary layer flows. Further, when $n_1 = 0$, the particles are not free to rotate near the surface. However, when $n_1 = 0.5$ or $n_1 = 1.0$, the microrotation term gets augmented and induces flow enhancement.

From the Eq. (1), it is clear that the suction velocity normal to the plate is a function of time only, which is considered to be in the form [1], [10] and [18]:

$$v^* = -(1 + \varepsilon A e n^* t^*) V_0, \tag{7}$$

where A is a real positive constant, εA small less than unity and V_0 is a scale of suction velocity which has a non-zero positive constant.

The radiative heat flux term by using the Rosseland approximation [14] is given by

$$q_r^* = -\frac{4\sigma^*}{3K_1^*}\frac{\partial T^{*4}}{\partial y^*}.$$

We assume that the temperature difference within the flow are sufficiently small such that T^{*4} may be expressed as a linear function of the temperature. This is accomplished by expanding T^{*4} in a Taylor series about T_∞^* and neglecting higher (second order onwards) order term [12], we get

$$T^{*4} \cong 4T_{\infty}^{*3}T^* - 3T_{\infty}^{*4}.$$

It is convenient to employ the following dimensionless variables:

$$u^* = U_0 u,\quad v^* = V_0 v,\quad y^* = \frac{v}{V_0}y,\quad u_p{}^* = U_0 U_p,\quad \omega^* = \frac{U_0 V_0}{v}\omega,$$

$$t^* = \frac{v}{V_0{}^2}t,\quad T - T_\infty = (T_\omega - T_\infty)\theta,\quad C - C_\infty = (C_\omega - C_\infty)\phi,\quad n^* = \frac{V_0{}^2}{v}n,$$

$$j^* = \frac{v^2}{V_0{}^2}j,\quad Pr = \frac{v}{\alpha},\quad Sc = \frac{v}{D},\quad M = \frac{\sigma B_0{}^2 v}{\rho V_0{}^2},\quad G_{rT} = \frac{vg\beta_T(T_\omega - T_\infty)}{U_0 V_0{}^2},$$

(8)

$$G_{rC} = \frac{vg\beta_C(C_\omega - C_\infty)}{U_0 V_0{}^2},\quad \gamma = \left(\mu + \frac{\Lambda}{2}\right)j^* = \mu j^*\left(1 + \frac{\beta}{2}\right),$$

$$\beta = \frac{\Lambda}{\mu} = \frac{v_r}{v},\quad K_2 = \frac{KU_0V_0{}^2}{v^2},\quad \eta = \frac{\mu j^*}{\gamma} = \frac{2}{2+\beta},\quad \gamma_1 = \frac{v\gamma_1{}^*}{VV_0{}^2},$$

$$E_c = \frac{U_0^2}{c_p(T_\omega - T_\infty)},\quad Pr = \frac{\mu c_p}{K},\quad v = \frac{\mu}{\rho},\quad Nr = \frac{4\sigma^* T_\infty^{*3}}{kK_1^*}.$$

With the help of Eq. (8), Eqs. (1), (2), (3), (4), (5), (6) and (7) reduced to the following:

$$\frac{\partial u}{\partial t} - (1 + \varepsilon A e^{nt})\frac{\partial u}{\partial y} = (1+\beta)\frac{\partial^2 u}{\partial y^2} + 2\beta\frac{\partial \omega}{\partial y} + G_{rT}\theta + G_{rC}\phi - Mu - \frac{1+\beta}{K_2}u,$$

(9)

$$\frac{\partial \omega}{\partial t} - (1 + \varepsilon A e^{nt})\frac{\partial \omega}{\partial y} = \frac{1}{\eta}\frac{\partial^2 \omega}{\partial y^2}, \qquad (10)$$

$$\frac{\partial \theta}{\partial t} - (1 + \varepsilon A e^{nt})\frac{\partial \theta}{\partial y} = \frac{1}{Pr}\left(1 + \frac{4Nr}{3}\right)\frac{\partial^2 \theta}{\partial y^2} + E_c\left(\frac{\partial u}{\partial y}\right)^2, \qquad (11)$$

$$\frac{\partial \phi}{\partial t}-(1+\varepsilon A e^{nt})\frac{\partial \phi}{\partial y}=\frac{1}{Sc}\frac{\partial^2 \phi}{\partial y^2}+\gamma_1\phi, \tag{12}$$

with the following dimensionless boundary conditions

$$u=U_p, \quad \omega=-n_1\frac{\partial u}{\partial y}, \quad \theta=1+\varepsilon e^{nt}, \quad \phi=1+\varepsilon e^{nt} \quad at\ y=0,$$

$$u\to 0, \quad \omega\to 0, \quad \theta\to 0, \quad \phi\to 0 \quad \text{as } y\to\infty. \tag{13}$$

To solve Eqs. (9), (10), (11) and (12) subject to the boundary conditions Eq. (13) we may use the following linear transformations for low value of ε [18] and [26]:

$$u(y,t)=u_0(y)+\varepsilon e^{nt}u_1(y)+O(\varepsilon^2),$$

$$\omega(y,t)=\omega_0(y)+\varepsilon e^{nt}\omega_1(y)+O(\varepsilon^2),$$

$$\theta(y,t)=\theta_0(y)+\varepsilon e^{nt}\theta_1(y)+O(\varepsilon^2), \tag{14}$$

$$\phi(y,t)=\phi_0(y)+\varepsilon e^{nt}\phi_1(y)+O(\varepsilon^2).$$

After substituting Eq. (14) into Eqs. (9), (10), (11), (12) and (13), we have

$$(1+\beta)u_0''+u_0'-\left(M+\frac{1+\beta}{K_2}\right)u_0=-G_{rT}\theta_0-G_{rC}\phi_0-2\beta\omega_0', \tag{15}$$

$$(1+\beta)u_1''+u_1'-\left(M+\frac{1+\beta}{K_2}-n\right)u_1=-Au_0'-G_{rT}\theta_1-G_{rC}\phi_1-2\beta\omega_1', \tag{16}$$

$$\omega''_0+\eta\omega'_0=0, \tag{17}$$

$$\omega''_1+\eta\omega'_1-n\eta\omega_1=-A\eta\omega'_0, \tag{18}$$

$$(3+4Nr)\theta''_0+3Pr\theta'_0+3PrE_cu'^2_0=0, \tag{19}$$

$$(3+4Nr)\theta''_1+3Pr\theta'_1-3nPr\theta_1=-3APr\theta'_0-6PrE_cu'_0u'_1, \tag{20}$$

$$\phi''_0+Sc\phi'_0+\gamma_1Sc\phi_0=0, \tag{21}$$

$$\phi''_1+Sc\phi'_1+(\gamma_1-n)Sc\phi_1=-ASc\phi'_0, \tag{22}$$

with the following boundary conditions:

$$u_0=U_p,\quad u_1=0,\quad \omega_0=-n_1u'_0,\quad \omega_1=-n_1u'_1,$$

$$\theta_0=1,\quad \theta_1=1,\quad \phi_0=1,\quad \phi_1=1\quad \text{at } y=0,$$

$$u_0=u_1=\omega_0=\omega_1=\theta_0=\theta_1=\phi_0=\phi_1=0\quad \text{as } y\to\infty. \tag{23}$$

To solve the nonlinear coupled Eqs. (15), (16),c (17), (18), (19), (20), (21) and (22), we assume that the viscous dissipation parameter (Eckert number E_c) is small, so we can write the asymptotic expansion as follows

$$u_0(y) = u_{01}(y) + E_c u_{02}(y) + O(E_c^2),$$

$$u_1(y) = u_{11}(y) + E_c u_{12}(y) + O(E_c^2),$$

$$\omega_0(y) = \omega_{01}(y) + E_c \omega_{02}(y) + O(E_c^2),$$

$$\omega_1(y) = \omega_{11}(y) + E_c \omega_{12}(y) + O(E_c^2),$$

$$\theta_0(y) = \theta_{01}(y) + E_c \theta_{02}(y) + O(E_c^2), \qquad (24)$$

$$\theta_1(y) = \theta_{11}(y) + E_c \theta_{12}(y) + O(E_c^2),$$

$$\phi_0(y) = \phi_{01}(y) + E_c \phi_{02}(y) + O(E_c^2),$$

$$\phi_1(y) = \phi_{11}(y) + E_c \phi_{12}(y) + O(E_c^2),$$

Substituting Eq. (24) into Eqs. (15), (16), (17), (18), (19), (20), (21) and (22), we obtain the following sequence of approximations for O(0) of E_c:

$$(1+\beta)u''_{01}+u'_{01}-\left(M+\frac{1+\beta}{K_2}\right)u_{01}=-G_{rT}\theta_{01}-G_{rC}\phi_{01}-2\beta\omega'_{01}, \tag{25}$$

$$(1+\beta)u''_{02}+u'_{02}-\left(M+\frac{1+\beta}{K_2}\right)u_{02}=-G_{rT}\theta_{02}-G_{rC}\phi_{02}-2\beta\omega'_{02}, \tag{26}$$

$$\omega''_{01}+\eta\omega'_{01}=0, \tag{27}$$

$$\omega''_{02}+\eta\omega'_{02}=0, \tag{28}$$

$$(3+4Nr)\theta''_{01}+3Pr\theta'_{01}=0, \tag{29}$$

$$(3+4Nr)\theta''_{02}+3Pr\theta'_{02}=-3Pru'^{2}_{01}, \tag{30}$$

$$\phi''_{01}+Sc\phi'_{01}+\gamma_1 Sc\phi_{01}=0, \tag{31}$$

$$\phi''_{02}+Sc\phi'_{02}+\gamma_1 Sc\phi_{02}=0, \tag{32}$$

subject to the boundary conditions:

$$u_{01}=U_p, \quad u_{02}=0, \quad \omega_{01}=-n_1u'_{01}, \quad \omega_{02}=-n_1u'_{02},$$

$\theta_{01}=1, \quad \theta_{02}=0, \quad \phi_{01}=1, \quad \phi_{02}=0 \quad \text{at } y=0,$

$$u_{01}=u_{02}=\omega_{01}=\omega_{02}=\theta_{01}=\theta_{02}=\phi_{01}=\phi_{02}=0 \quad \text{as } y\rightarrow\infty. \tag{33}$$

Also we get the following equations for O(1) of E_c:

$$(1+\beta)u_{11}''+u_{11}'-\left(M+\frac{1+\beta}{K_2}-n\right)u_{11}=-Au_{01}'-G_{rT}\theta_{11}-G_{rC}\phi_{11}-2\beta\omega_{11}', \tag{34}$$

$$(1+\beta)u_{12}''+u_{12}'-\left(M+\frac{1+\beta}{K_2}-n\right)u_{12}=-Au_{02}'-G_{rT}\theta_{12}-G_{rC}\phi_{12}-2\beta\omega_{12}', \tag{35}$$

$$\omega_{11}''+\eta\omega_{11}'-n\eta\omega_{11}=-A\eta\omega_{01}', \tag{36}$$

$$\omega_{12}''+\eta\omega_{12}'-n\eta\omega_{12}=-A\eta\omega_{02}', \tag{37}$$

$$(3+4Nr)\theta_{11}''+3Pr\theta_{11}'-3nPr\theta_{11}=-3APr\theta_{01}', \tag{38}$$

$$(3+4Nr)\theta_{12}''+3Pr\theta_{12}'-3nPr\theta_{12}=-3APr\theta_{02}'-6Pru_{01}'u_{11}', \tag{39}$$

$$\phi''_{11}+Sc\phi'_{11}+(\gamma_1-n)Sc\phi_{11}=-ASc\phi'_{01}, \tag{40}$$

$$\phi''_{12}+Sc\phi'_{12}+(\gamma_1-n)Sc\phi_{12}=-ASc\phi'_{02}, \tag{41}$$

subject to the boundary conditions:

$$u_{11}=0,\quad u_{12}=0,\quad \omega_{11}=-n_1u'_{11},\quad \omega_{12}=-n_1u'_{12},$$

$$\theta_{11}=1,\quad \theta_{12}=0,\quad \phi_{11}=1,\quad \phi_{12}=0 \quad \text{at } y=0,$$

$$u_{11}=u_{12}=\omega_{11}=\omega_{12}=\theta_{11}=\theta_{12}=\phi_{11}=\phi_{12}=0 \quad \text{as } y\to\infty. \tag{42}$$

Solving Eqs. (25), (26), (27), (28), (29), (30), (31) and (32) under the boundary conditions Eq. (33) and Eqs. (34), (35), (36), (37), (38), (39), (40) and (41) under the boundary conditions Eq. (42) and substituting into Eqs. (24) and (14), we obtain the temperature, angular velocity and velocity profiles of all the flow respectively as follows:

$$\begin{aligned}
u(y,t)=\,& a_1e^{-h_2y}+a_2e^{-h_1y}+a_3e^{-\eta y}+a_4e^{-h_3y}+E_c(a_{16}e^{-h_2y}+a_{17}e^{-2h_3y}\\
&+a_{18}e^{-2h_2y}+a_{19}e^{-(h_2+h_3)y}+a_{20}e^{-2h_1y}+a_{21}e^{-2\eta y}+a_{22}e^{-(h_1+\eta)y}\\
&+a_{23}e^{-(h_1+h_3)y}+a_{24}e^{-(h_3+\eta)y}+a_{25}e^{-(h_1+h_2)y}+a_{26}e^{-(h_2+\eta)y}\\
&+a_{27}e^{-\eta y}+a_{28}e^{-h_3y})+\varepsilon e^{nt}(b_5e^{-h_3y}+b_6e^{-h_2y}+b_7e^{-h_1y}+b_8e^{-\eta y}\\
&+b_9e^{-h_5y}+b_{10}e^{-h_2y}+b_{11}e^{-h_4y}+b_{12}e^{-h_1y}+b_{13}e^{-h_6y}+b_{14}e^{-\eta y}\\
&+b_{15}e^{-h_7y}+E_c(e_1e^{-h_3y}+e_2e^{-h_5y}+e_3e^{-(h_3+h_7)y}+e_4e^{-(h_3+h_5)y}\\
&+e_5e^{-(h_3+h_4)y}+e_6e^{-(h_3+h_6)y}+e_7e^{-(h_2+h_7)y}+e_8e^{-(h_2+h_5)y}+e_9e^{-(h_2+h_4)y}\\
&+e_{10}e^{-(h_2+h_6)y}+e_{11}e^{-(h_1+h_7)y}+e_{12}e^{-(h_1+h_5)y}+e_{13}e^{-(h_1+h_4)y}\\
&+e_{14}e^{-(h_1+h_6)y}+e_{15}e^{-(h_7+\eta)y}+e_{16}e^{-(h_5+\eta)y}+e_{17}e^{-(h_4+\eta)y}\\
&+e_{18}e^{-(h_6+\eta)y}+e_{19}e^{-h_6y}+e_{20}e^{-h_7y}+d_1e^{-h_2y}+d_2e^{-2h_3y}+d_3e^{-2h_2y}\\
&+d_4e^{-2h_1y}+d_5e^{-2\eta y}+d_6e^{-\eta y}+d_7e^{-(h_2+h_3)y}+d_8e^{-(h_1+\eta)y}\\
&+d_9e^{-(h_1+h_3)y}+d_{10}e^{-(h_1+h_2)y}+d_{11}e^{-(h_2+\eta)y}+d_{12}e^{-(h_3+\eta)y}))
\end{aligned}$$

$$\omega(y,t) = a_{29}e^{-\eta y} + E_c a_{30}e^{-\eta y} + \varepsilon e^{nt}(e_{21}e^{-h_6 y} + b_3 e^{-\eta y} + E_c(e_{22}e^{-h_6 y} + b_4 e^{-\eta y}))$$

$$\begin{aligned}\theta(y,t) = {} & e^{-h_2 y} + E_c(a_5 e^{-2h_3 y} + a_6 e^{-2h_2 y} + a_7 e^{-(h_2+h_3)y} + a_8 e^{-2h_1 y} + a_9 e^{-2\eta y} \\ & + a_{10}e^{-(h_1+\eta)y} + a_{11}e^{-(h_1+h_3)y} + a_{12}e^{-(h_3+\eta)y} + a_{13}e^{-(h_1+h_2)y} \\ & + a_{14}e^{-(h_2+\eta)y} + a_{15}e^{-h_2 y}) + \varepsilon e^{nt}((1-b_1)e^{-h_5)y} + b_2 e^{-h_2)y} \\ & + E_c(b_{16}e^{-h_2 y} + b_{27}e^{-(h_3+h_7)y} + b_{32}e^{-(h_3+h_5)y} + b_{34}e^{-(h_3+h_4)y} \\ & + b_{36}e^{-(h_3+h_6)y} + b_{38}e^{-(h_2+h_7)y} + b_{43}e^{-(h_2+h_5)y} + b_{45}e^{-(h_2+h_4)y} \\ & + b_{47}e^{-(h_2+h_6)y} + b_{49}e^{-(h_2+h_7)y} + b_{54}e^{-(h_1+h_5)y} + b_{56}e^{-(h_1+h_4)y} \\ & + b_{58}e^{-(h_1+h_6)y} + b_{60}e^{-(h_7+\eta)y} + b_{65}e^{-(h_5+\eta)y} + b_{67}e^{-(h_4+\eta)y} \\ & + b_{69}e^{-(h_6+\eta)y} + b_{71}e^{-h_5 y} + d_{13}e^{-2h_2 y} + d_{14}e^{-2h_1 y} + d_{15}e^{-2h_3 y} \\ & + d_{16}e^{-2\eta y} + d_{17}e^{-(h_1+\eta)y} + d_{18}e^{-(h_2+\eta)y} + d_{19}e^{-(h_3+\eta)y} \\ & + d_{20}e^{-(h_2+h_3)y} + d_{21}e^{-(h_1+h_3)y} + d_{22}e^{-(h_1+h_2)y}))\end{aligned}$$

$\phi(y, t)=e-h_1y+\varepsilon ent((1-b_1)e-h_4y+b_1e-h_1y)$

where a_i, b_i, c_i, d_i, and h_i are provided in the Appendix.

The local skin friction coefficient, local wall couple stress coefficient, local Nusselt number, and local Sherwood number are important physical quantities for this type of heat and mass transfer problem which are defined below.

The wall shear stress may be written as

$$\begin{aligned}\tau_\omega^* &= (\mu+\Lambda)\left.\frac{\partial u^*}{\partial y^*}\right|_{y^*=0} + \Lambda\omega^*\big|_{y^*=0} \\ &= \rho U_0 V_0[1+(1-n_1)\beta]u'(0)\end{aligned} \quad (43)$$

Therefore, the local skin-friction factor is given by

$$C_f = \frac{2\tau_\omega^*}{\rho U_0 V_0} = 2[1+(1-n_1)\beta]u'(0). \tag{44}$$

The wall couple stress may be written as:

$$M_\omega = \gamma \frac{\partial \omega^*}{\partial y^*}\bigg|_{y=0}. \tag{45}$$

Therefore, the local couple stress coefficient is given by

$$C'_\omega = \frac{M_\omega v^2}{\gamma U_0 V_0^2} = \omega'(0). \tag{46}$$

The rate of heat transfer at the surface in terms of the local Nusselt number can be written as:

$$Nu_x = x\frac{(\partial T/\partial y^*)_{y^*=0}}{T_\infty - T_\omega}. \tag{47}$$

Using Eqs. (8) and (45) in Eq. (47), we get

$$Nu_x Re_x^{-1} = -\theta'(0),$$

where $Re_x = \frac{xV_0}{v}$ is the local Reynolds number.

The rate of mass transfer at the surface in terms of the local Sherwood number is given by:

$$Sh_x = x\frac{(\partial C/\partial y^*)_{y^*=0}}{C_\infty - C_\omega}. \tag{48}$$

Using Eqs. (8) and (47) in Eq. (48), we get

$$Sh_x Re_x^{-1} = -\phi'(0).$$

3. RESULTS AND DISCUSSIONS

To assess the physical properties of the problem, the effects of various parameters like Eckert number, local Grashof number, permeability parameter, magnetic field parameter, Prandtl number, thermal radiation parameter, chemical reaction parameter, viscosity parameter and Schmidt number, velocity component in *x*-direction, component of angular velocity, temperature of the fluid, concentration of species, skin friction coefficient, Nusselt number and Sherwood number are analyzed. Numerical evaluations of the analytical solutions were performed and the results are presented in graphical and tabular forms. The values of the physical parameters such as Nr, ω etc. are taken from the literature [2], [7] and [9]. The variation in velocity profile with y for various values in thermal Grashof numbers are shown in Fig. 1. This figure reflects that with increase in G_{rT}there is increase in fluid velocity due to enhancement of the buoyancy force. The curves show that the velocity starts from a minimum value on the surface and increases till it attains a peak value near the plate, then it starts decreasing till the end of the boundary layer. The positive value of G_{rT} indicates the cooling of the plate and it is observed that velocity increases rapidly near the wall of the plate and then decays to the free stream velocity. When G_{rT} is negative, the plate becomes hotter and there is retardation in the fluid velocity. Thus the thermal Grashof number G_{rT} signifies the relative effect of the thermal buoyancy force to the viscous hydrodynamic force in the boundary layer. As expected, it is observed that there is a rise in the velocity as G_{rT} increases. The solutal Grashof number G_{rC} defines the ratio of the species buoyancy force to the viscous hydrodynamic force. As expected, the fluid velocity increases as Grashof number G_{rC}increases as seen in Fig. 2.

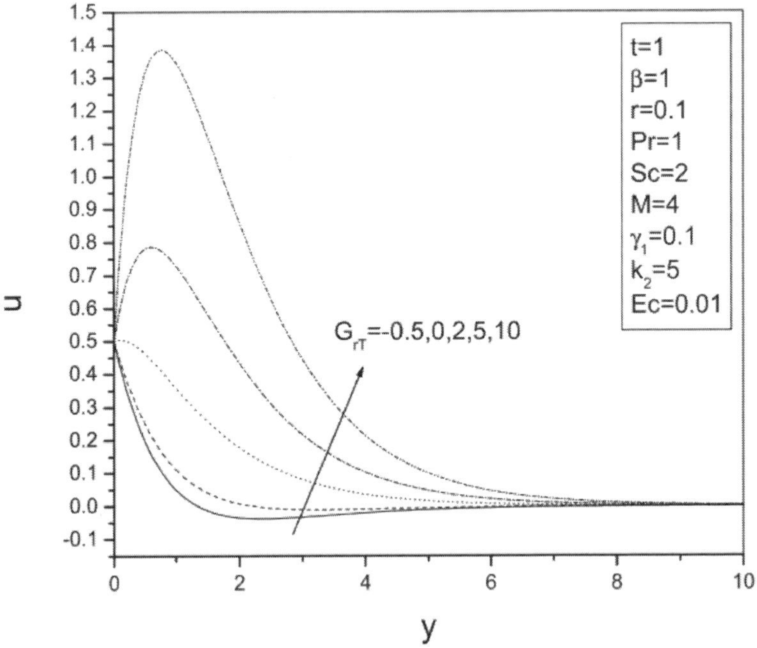

Figure 1. Velocity profiles vs. y for different values of G_{rT}.

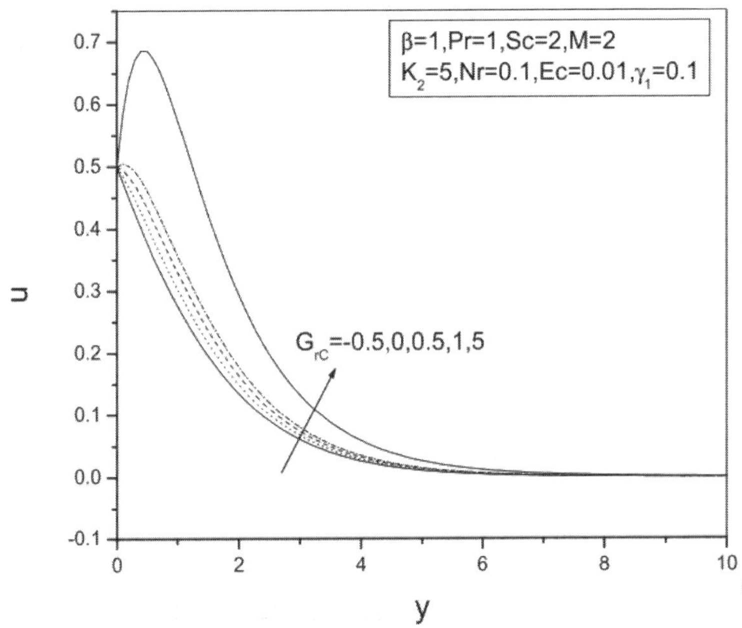

Figure 2. Velocity profiles vs. y for different values of G_{rC}.

The effects of first order irreversible chemical reaction parameter on the velocity profiles are studied in Fig. 3. It is seen that the velocity profile increases with an increase in the chemical reaction parameter. Also, it is observed that when the chemical reaction parameter is negative (i.e. generative reaction) there is an increase in the fluid flow velocity. In fact, as chemical reaction increases there is a considerable amount of reduction in the velocity profiles, and the presence of the peak indicates that the maximum value of the velocity occurs in the fluid close to the surface but not at the surface. For positive chemical reaction parameter (i.e. destructive reaction), there is reduction in the fluid flow velocity. Fig. 4 depicts the variation of the microrotational velocity distribution with permeability parameter of the porous medium for different time t. It is clearly observed that there is a decrease in the microrotational velocity distribution with an increase in the value of the permeability parameter. The physical reason behind this decrease is due to the fact that velocity of the fluid dominates over microrotational velocity distribution with an increase in the permeability parameter. Thus the higher the value of the porous permeability the lower the microrotational velocity, which results in increasing the velocity of the fluid.

It is observed from Fig. 5 that the decrease in the value of ω is very rapid for the higher value of Eckert number. It is to be noted that the parameter n_1 is associated with the boundary condition Eq. (23) and that it physically reduces to the concentration of the microelement at the plate. Further, it is noticed that this initial boundary condition for microrotational corresponds to the vanishing of the anti-symmetric part of the stress tensor and related to the weak concentration of the micro-elements of the micropolar fluid. Thus it is clear that greater n_1 decreases the value of the microrotational velocity distribution rapidly. The effects of micro-rotational velocity distribution against y for different value of Eckert number are shown in Fig. 6. This figure depicts that the micro-rotational velocity distribution decreases with an increase in the Eckert number whereas its value increases with an increase in y. Physically, this is due to the fact that increase in the Eckert number increases dissipation of energy by viscous force and hence thereby decreases the micro-rotational velocity by increasing the Eckert number.

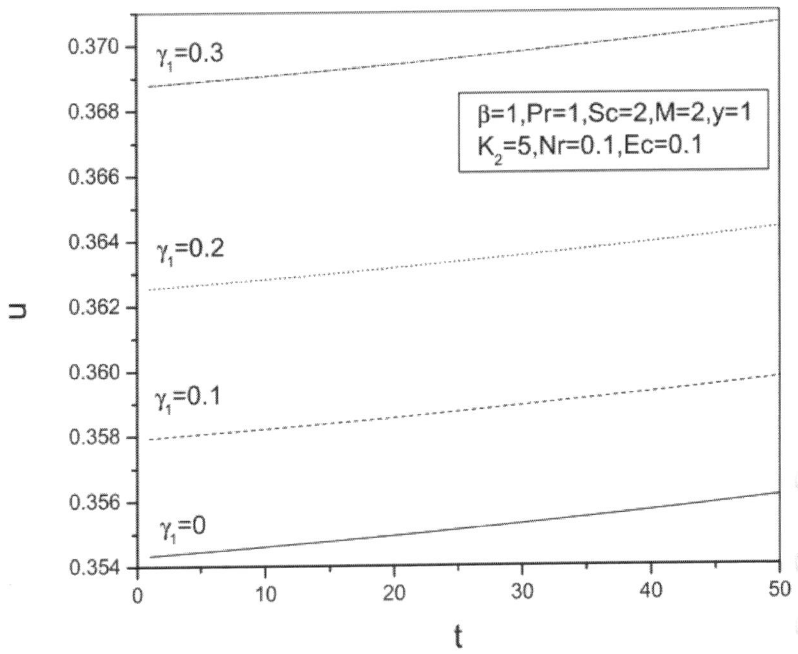

Figure 3. Velocity profiles vs. t for different values of chemical reaction parameter.

Fig. 7 depicts the variation of micro-rotational velocity distribution against y for different values of the magnetic field parameter. Here, we observed that the micro-rotational velocity distribution increases with an increase in the magnetic field parameter and it becomes zero as y tends to ∞ for all the values of the magnetic field parameter. Further, it is observed that for higher values of y, all the curves of ω coincide at that point (i.e. $y = 8$). Physically, an increase in the strength of the magnetic field enhances the microrotational velocity. The variation of microrotational velocity distribution for various values of the chemical reaction parameter are shown in Fig. 8. From this figure we see that microrotational velocity distribution decreases by increasing the value of the chemical reaction parameter. The effect of permeability parameter on temperature profile is shown in Fig. 9 with time. From this figure, it is observed that the temperature increases with an increase in the permeability parameter K_2 for all time t. This is due to the fact that as the velocity increases there is an increase in the viscous dissipation as it adds energy to the fluid and hence temperature of the fluid increases with the permeability parameter K_2. Also, the effects of G_{rT} on temperature profile are shown in Fig. 10. For the positive value of G_{rT}, i.e., for externally cooled plate the temperature increases, but for the negative value of the G_{rT}, i.e., for externally heated plate, the temperature reduces.

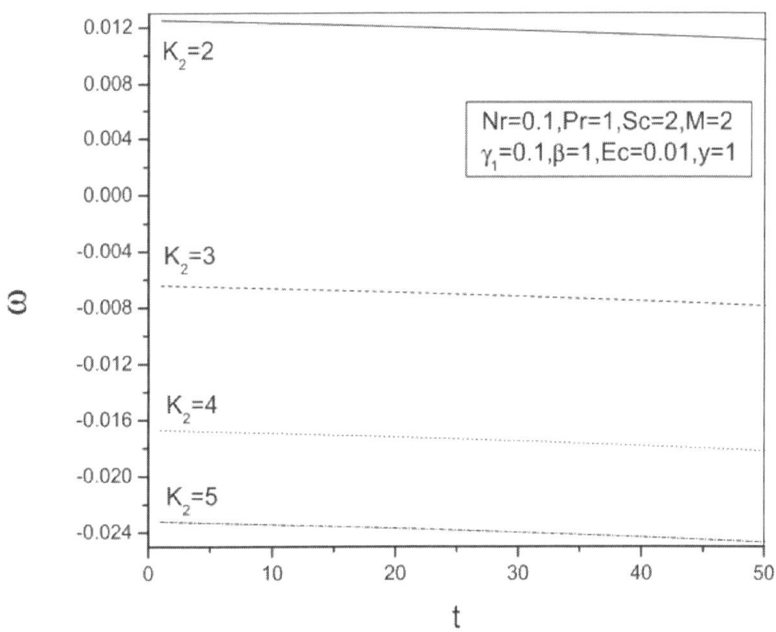

Figure 4. Plot of micro-rotational velocity distribution with time t for different values of K_2.

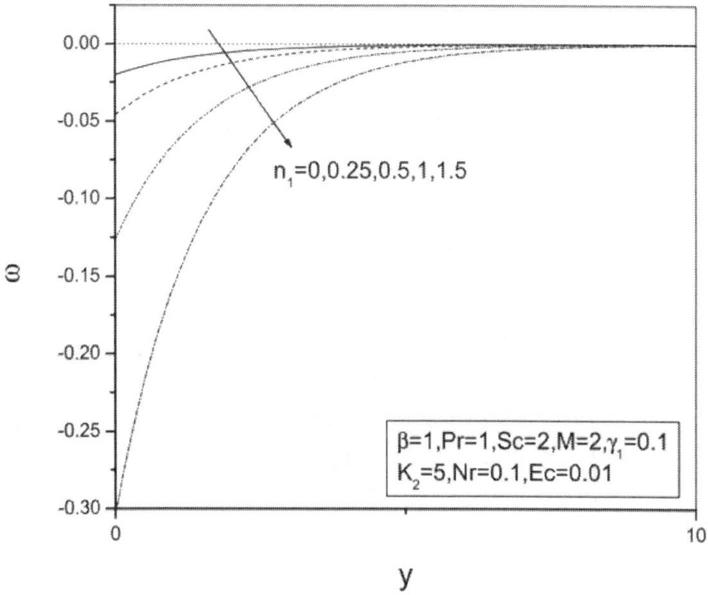

Figure 5. Plot of micro-rotational velocity distribution with y for different values of n_1.

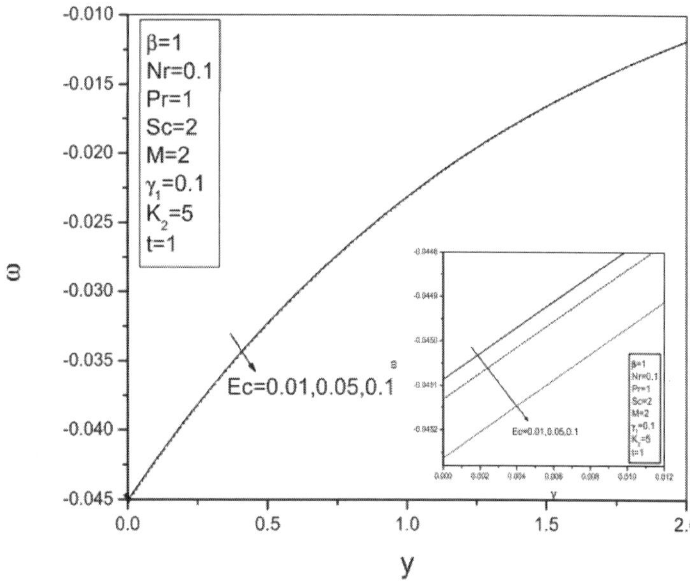

Figure 6. Microrotational velocity distribution vs. y for different values of Eckert number Ec.

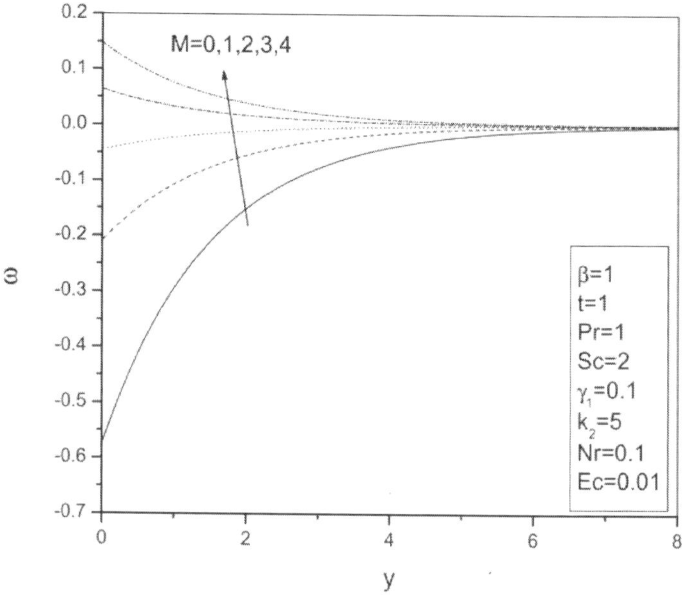

Figure 7. Microrotational velocity profile vs. y for different values of magnetic field parameter M.

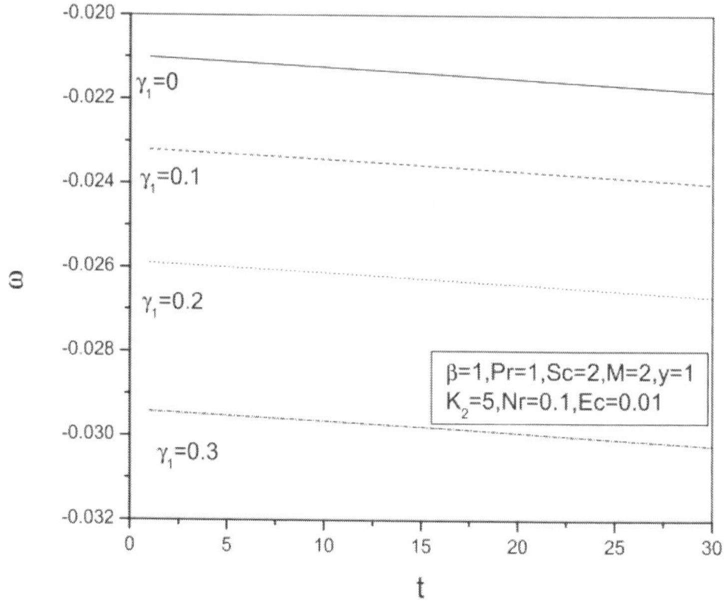

Figure 8. Micro-rotational velocity distribution vs. time t for different values of chemical reaction parameter.

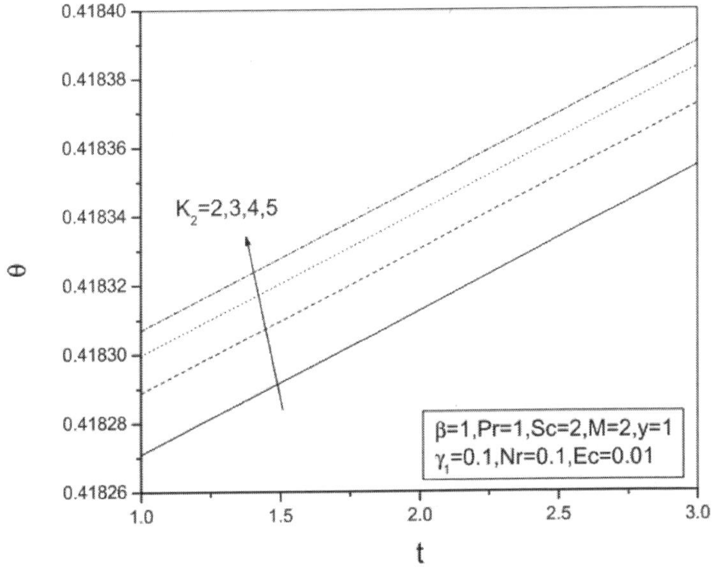

Figure 9. Variation of temperature profiles with time t different value of K_2.

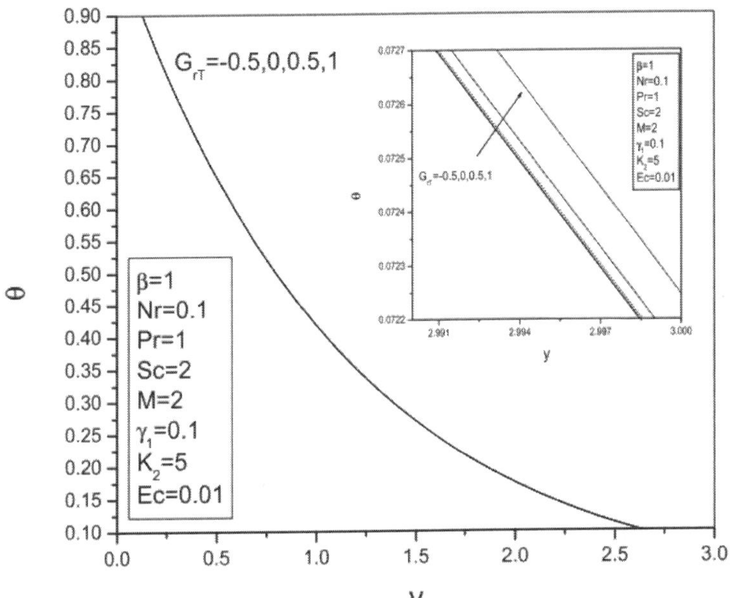

Figure 10. Variation of temperature profiles with y for different values of G_{rT}.

Fig. 11 illustrates the effect of thermal radiation on the temperature distribution with y for different values of the radiation parameter. For larger thermal radiation parameter the thermal boundary layer is thicker, so the temperature profile increases. It is also seen from this figure that the temperature distribution increases with an increase in the radiation parameter and decreases with an increase in y and the value reaches zero as $y \rightarrow \infty$ matching the boundary condition at that point. Fig. 12 is drawn to study the variations of temperature distribution against y for different value of Eckert number. In this figure we can see that the temperature distribution increases with an increase in the Eckert number and temperature decreases steeply with an increase in y. Physically, the Eckert number (*Ec*) expresses the relationship between the kinetic energy in the flow and the enthalpy. It embodies the conversion of kinetic energy into internal energy by work done against the viscous fluid stresses. Greater viscous dissipative heat causes a rise in the temperature as well as the velocity. Thus increase in the value of the Eckert number indicates that there is an increase in the viscous dissipation and thereby increases the temperature of the fluid as the dissipative force adds energy to the fluid.

Fig. 13 depicts the variation of temperature profiles for different values of the chemical reaction parameter. It is seen from this figure that the temperature profile increases with an increase in the chemical reaction parameter. The effect of skin-frictional coefficient against t for different value of chemical reaction parameter are depicted in Fig. 14. This figure shows that the local skin-frictional coefficient increases with increases in the chemical reaction parameter.

Fig. 15 depicts the variation of skin-frictional coefficient against time t for different value of the Eckert number. From this figure, it is observed that the skin-frictional coefficient increases with an increase in the Eckert number due to increase in the viscous dissipative forces. Fig. 16 depicts the variation of skin-frictional coefficient against t for different values of magnetic field parameter. It is seen from this figure that skin-frictional coefficient decreases with an increase in the magnetic field parameter. Also, it is observed from this figure that there is an increase in the value of the skin-friction coefficient with an increase in time and the rate of increase become rapid for large values of t. The profiles of the local couple stress coefficient at the wall against time t for different value of Eckert number are shown in Fig. 17. It is seen from this figure that by increasing the Eckert number there is remarkable increase in the local couple stress coefficient at the wall due to increase in the dissipative forces exerted by viscosity of the fluid.

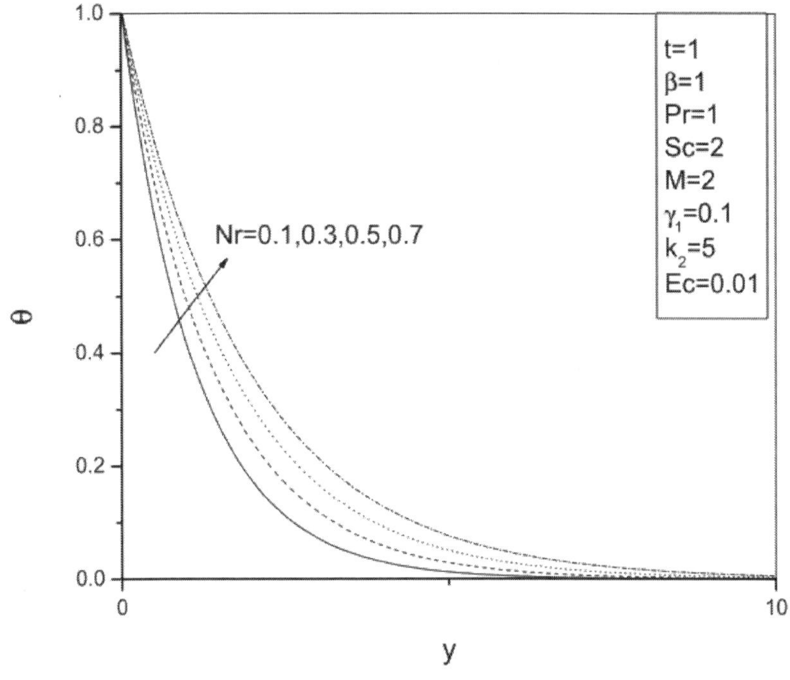

Figure 11. Variation of temperature profiles with y for different value of radiation parameter Nr.

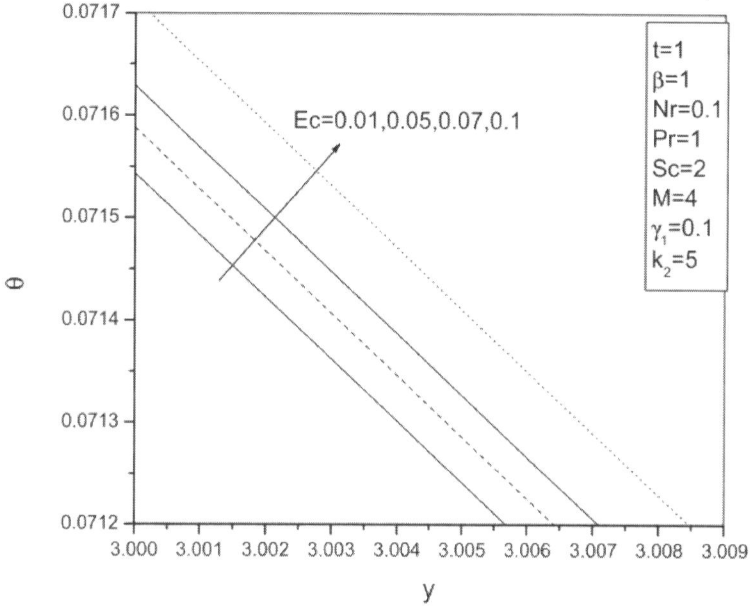

Figure 12. Temperature profile vs. y for different values of Eckert number Ec.

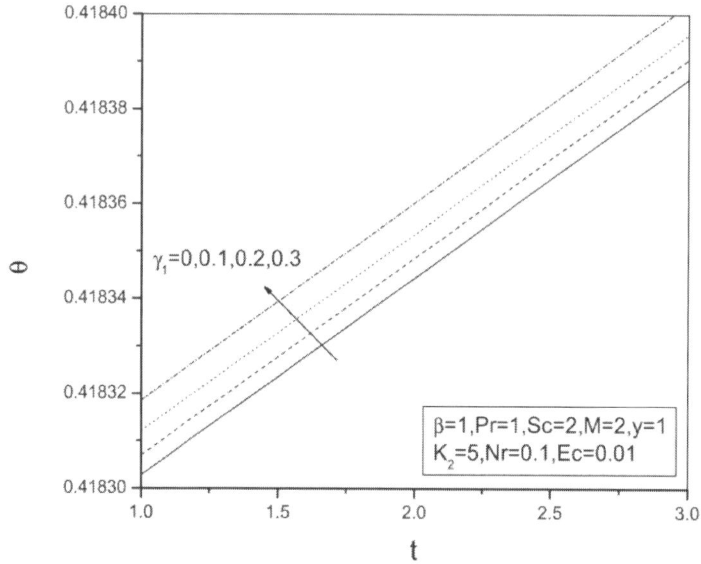

Figure 13. Variation of temperature profile with t for different value of chemical reaction parameter.

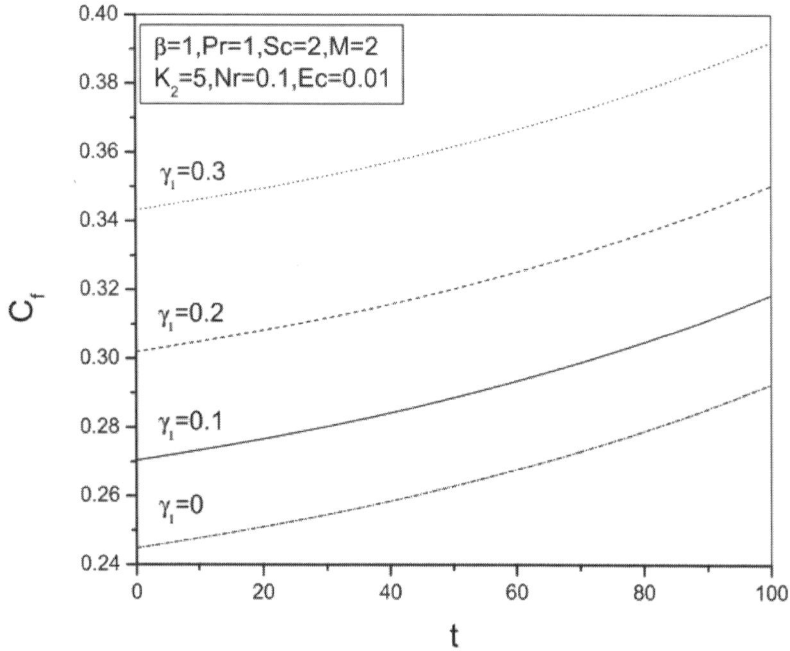

Figure 14. Variation of C_f with t for different values of chemical reaction parameter.

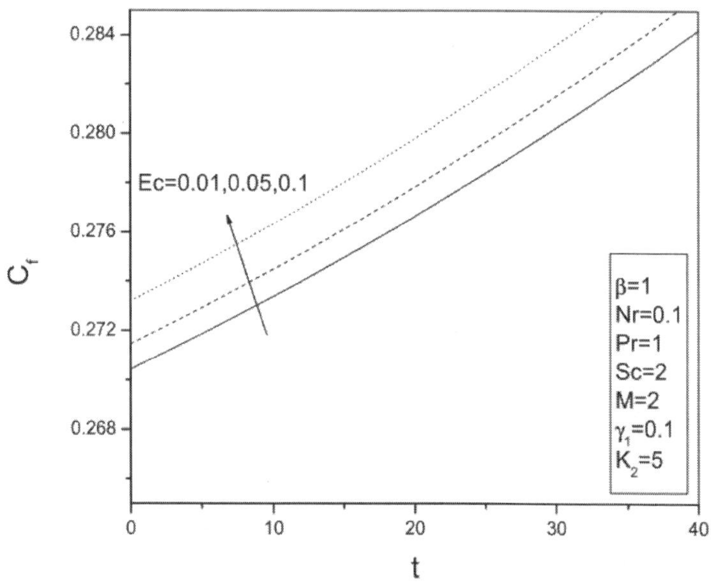

Figure 15. Variation of C_f with t for different values of Eckert number Ec.

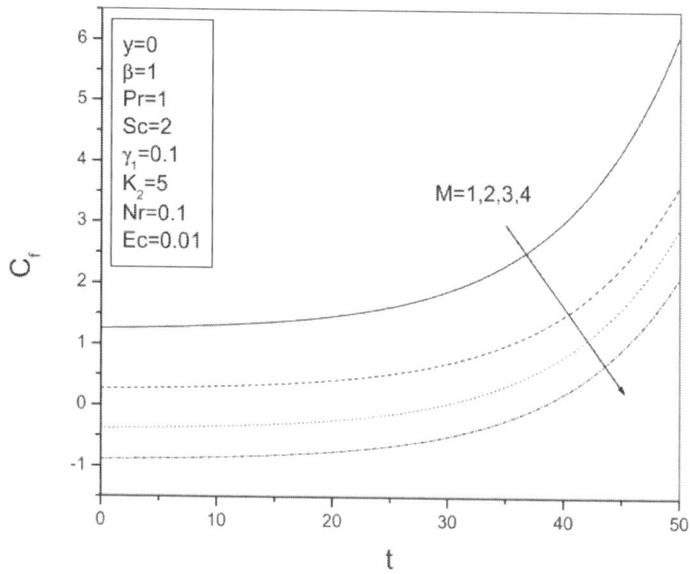

Figure 16. Variation of C_f with t for different values of M.

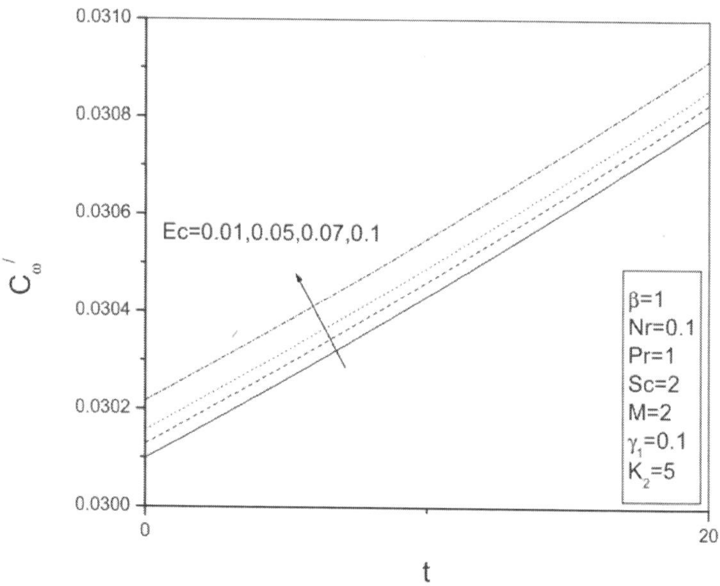

Figure 17. Variation of C'_{ω} with t for different values of Eckert number.

The effect of local couple stress coefficient at the wall against t for different values of the permeability parameter are shown in Fig. 18. It is seen from this figure that increase in the permeability parameter increases the local couple stress coefficient at the wall due to increase in the porosity of the porous medium. Fig. 19 depicts the variation of concentration at the wall against t for different value of magnetic field parameter. In this figure it is observed that the local couple stress coefficient at the wall decreases by increasing the magnetic field parameter and it rapidly decreases for large values of t. This causes the local couple stress coefficient buoyancy effects to decrease, yielding a reduction in the fluid velocity. The reduction in the concentration profiles is accompanied by simultaneous reduction in the momentum and concentration boundary layers thickness.

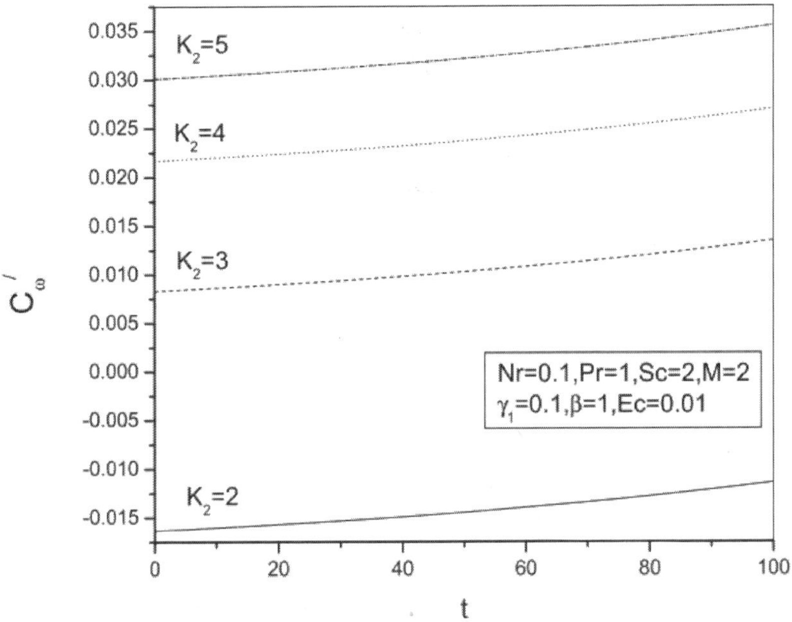

Figure 18. Variation of C'_ω with time for different values of K_2.

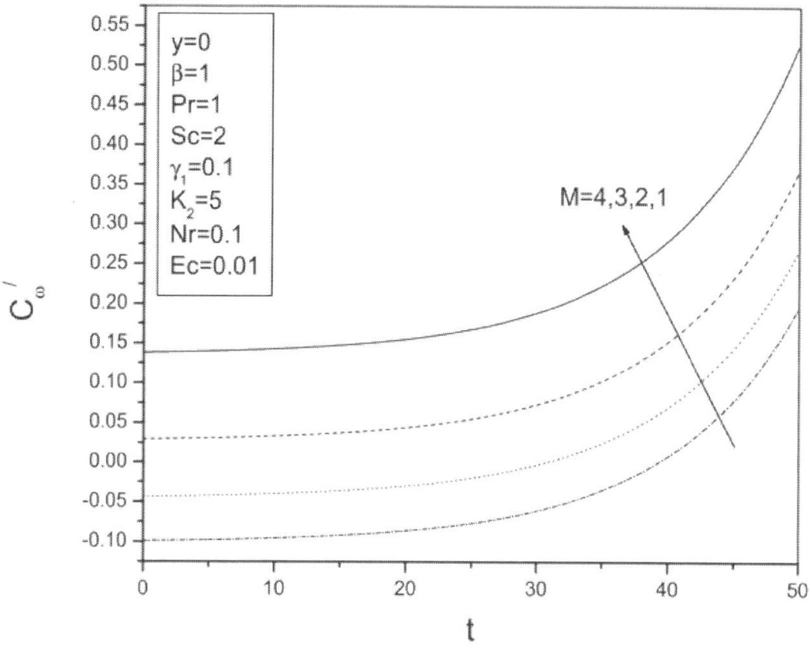

Figure 19. Plot of C'_ω with t for different values of M.

Fig. 20 depicts the variation of Nusselt number against t for different values of the Eckert number. We observe that the Nusselt number decreases with an increase in the Eckert number due to the fact that increasing the values of the Eckert number generates heat in the fluid due to frictional heating. Thus the effect of increasing *Ec* is to enhance the temperature and thereby increases the Nusselt number. The plot of Nusselt number against time t for different value of magnetic field parameter is shown in Fig. 21. It is seen from this figure that the local Nusselt number increases significantly for small values of the magnetic field parameter and the effect become insignificant for large values in the magnetic field parameter.

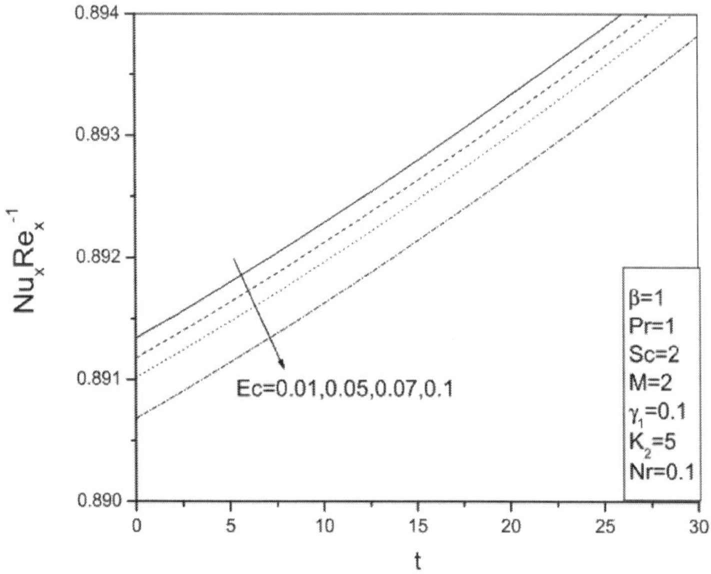

Figure 20. Plot of $Nu_xRe_x^{-1}$ with time for different values of Eckert number.

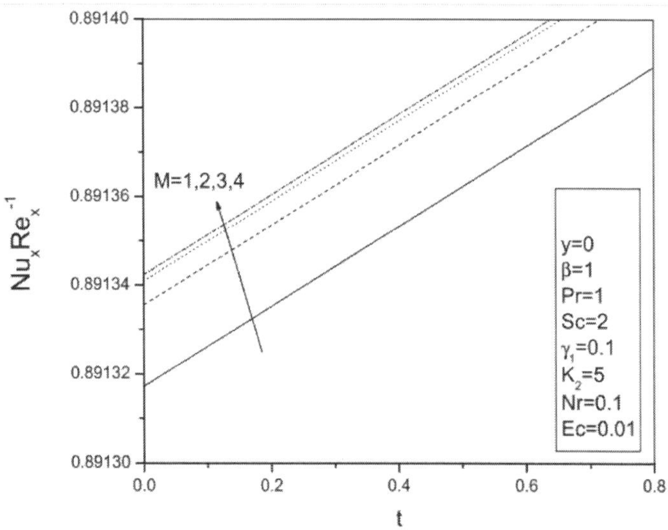

Figure 21. Plot of $Nu_xRe_x^{-1}$ with time for different values of M.

Fig. 22 depicts the variation of Sherwood number against t for different values of the Schmidt number. It is well known that the Schmidt number Sc embodies the ratio of the momentum to the mass diffusivity. The Schmidt number which quantifies the relative effectiveness of momentum and mass transport by diffusion in the hydrodynamic (velocity) and concentration (species) boundary layers. As the Schmidt number increases the concentration decreases. This causes the concentration buoyancy effects to decrease, yielding a reduction in the fluid velocity. The reduction in the velocity and concentration profiles is accompanied by simultaneous reduction in the velocity and concentration boundary layers. Thus it is observed from this figure that the Sherwood number increases by increasing the Schmidt number.

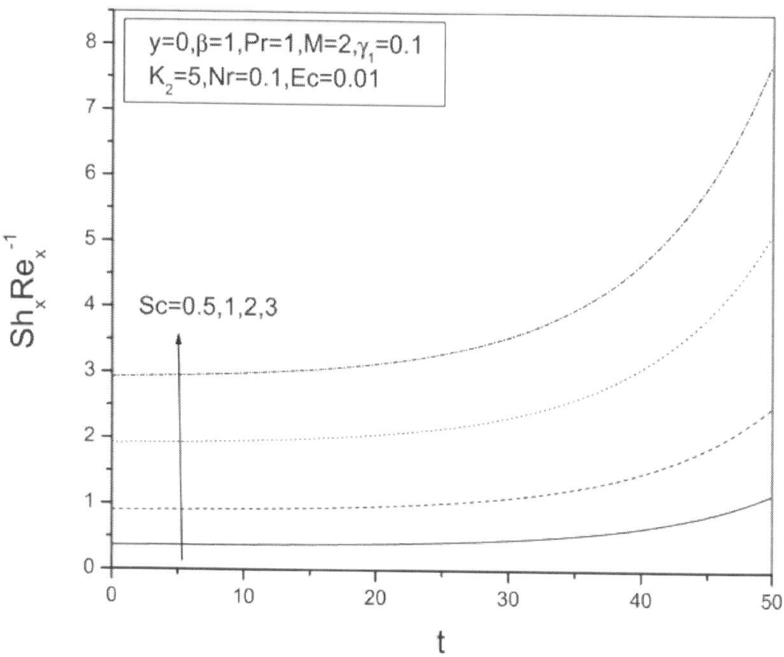

Figure 22. Plot of $Sh_x Re_x^{-1}$ with t for different values of Sc.

We have compared the result of Nusselt number and Sherwood number for different values of t with Modather et al. [7] in Table 1, in the absence of thermal radiation parameter and viscous dissipation effects. Numerical values of local skin friction factor, local couple stress coefficient at the wall, Nusselt number and Sherwood number for different values of the parameters *Sc*, *Pr*, K_2, *M*, *Nr* and E_c are shown in Table 2 by keeping the values of other parameters constant. We have taken the values of the parameter $\varepsilon = 0.01$, $n = 0.1$, $n_1 = 0.5$, $G_{rT} = 2$, $G_{rC} = 1$, $U_p = 0.5$, $A = 0.1$, for all the figures and tables presented in this paper. It is observed from this table that the effect of increase in the value of *Sc*, *Pr*, and *M* is to decrease C_f and C'_w whereas reverse effect is found for $Nu_x Re_x^{-1}$. Further, it is noticed that the effect of increase in the value of K_2, *Nr*, and *Ec* is to increase C_f and C'_w whereas reverse effect is seen for $Nu_x Re_x^{-1}$. Also, the effect of increasing the value of Schmidt number *Sc* is found to be very prominent on Sherwood number $Sh_x Re_x^{-1}$, i.e. an increasing trend is found on $Sh_x Re_x^{-1}$ by increasing the value of the Schmidt number.

Table 1. Comparison of the present result of $Nu_x Re_x^{-1}$ and $Sh_x Re_x^{-1}$ with Modather et al. [7] for different values of t when Pr=1.0, Sc=2.0, Nr=0.0, Ec=0.0, A=0.0, β=1.0, M=2.0, ε=0.01, n_1=0.5, n=0. 1, Gr_T=2, Gr_C=1, K_2=5, γ_1=0.1, Up=0.5.

	Modather et al. [7] results		Present results	
t	$Nu_x Re_x^{-1}$	$Sh_x Re_x^{-1}$	$Nu_x Re_x^{-1}$	$Sh_x Re_x^{-1}$
0	1.00887	1.91217	1.0088730	1.9121732
1	1.00981	1.91404	1.0098062	1.9140395
3	1.43567	1.91838	1.0119773	1.9183817
5	1.01463	1.92369	1.0146291	1.9236853
10	1.02412	1.94267	1.0241193	1.9426657
20	1.06556	2.02555	1.0655630	2.0255532
30	1.17822	2.25086	1.1782186	2.2508645
40	1.48445	2.86332	1.4844484	2.8633242
50	2.31687	4.52816	2.3168676	4.5281625

Table 2. Effects of various values of fluid properties on the coefficients of skin friction coefficient, couple stress coefficient, Nusselt number and Sherwood number.

Sc	Pr	K_2	M	Nr	E_c	C_f	C_w'	$Nu_x Re_x^{-1}$	$Sh_x Re_x^{-1}$
2	1	5	2	0.1	0.001	0.2707133	0.0301309	0.8914258	1.9158002
1						0.6682574	0.0745723	0.8914186	0.8975272
0.5						1.1339602	0.1264654	0.8914083	0.3664133
2	3					–0.8333037	–0.0927355	2.6738615	1.9158002
	5					–1.2005236	–0.1335068	4.4563727	1.9158002
	1	3				0.0745329	0.0083268	0.8914279	1.9158002
		1				–0.6985520	–0.0776972	0.8914325	1.9158002
		5	3			–0.3851199	–0.0427542	0.8914313	1.9158002
			4	0.7		–0.8853874	–0.0985380	0.8914328	1.9158002
			2	0.3		0.5051470	0.0560755	0.7216151	1.9158002
				0.5		0.6920367	0.0770952	0.6061646	1.9158002
				0.1	0.05	0.2717460	0.0301607	0.8912639	1.9158002
					0.07	0.2723846	0.0301891	0.8911055	1.9158002
					0.1	0.2734951	0.0302488	0.8907710	1.9158002

4. CONCLUSIONS

In this paper the effects of chemical reaction and viscous dissipation on MHD free convection flow in the presence of heat source or sink and thermal radiation with variable surface temperature and concentration have been studied analytically. Perturbation method is employed to solve the governing equations of the flow. From the present investigation, the following conclusions have been drawn:

(i) The temperature distribution increases with an increase in the Eckert number, radiation parameter, thermal Grashof number and permeability parameter.

(ii) The micro-rotational velocity distribution decreases with an increase in chemical reaction parameter, permeability parameter and Eckert number but increases with the increase in the magnetic field parameter.

(iii) There are increases in the skin frictional coefficient with increases in the Eckert number and chemical reaction parameter.

(iv) When there is increase in the magnetic field parameter there is decrease in skin friction coefficient.

(v) The value of the Nusselt number decreases with an increase in the Eckert number, and increases with an increase in the magnetic field parameter.

(vi) When increasing the value of Schmidt number there is increase in the Sherwood number.

(vii) Local couple stress coefficient at the wall decreases with an increase in the values of the magnetic field parameter and it increases with an increase in Eckert number and permeability parameter in the specified range. Thus we can conclude that the physical parameters play an important role on the rate of heat and mass transfer over a vertical permeable plate embedded in a porous medium.

ACKNOWLEDGMENTS

We thank the reviewers for their constructive comments that have led to a definite improvement in the paper.

REFERENCES

1. F.S. Ibrahim, A.M. Elaiw, A.A. Bakr, Effect of the chemical reaction and radiation absorption on the unsteady MHD free convection flow past a semi infinite vertical permeable moving plate with heat source and suction, Commun. Nonlinear Sci. Numer. Simul. 13 (2008) 1056–1066.

2. M.Q. Al-Odat, T.A. Al-Azab, Influence of chemical reaction on transient MHD free convection over a moving vertical plate, Emirates J. Eng. Res. 12 (3) (2007) 15–21.

3. [3] M.A. Seddeek, A.A. Darwish, M.S. Abdelmeguid, Effects of chemical reaction and variable viscosity on hydromagnetic mixed convection heat and mass transfer for Hiemenz flow through porous media with radiation, Commun. Nonlinear Sci. Numer. Simul. 12 (2007) 195–213.

4. D. Pal, B. Talukdar, Perturbation analysis of unsteady magnetohydrodynamic convective heat and mass transfer in a boundary layer slip flow past a vertical permeable plate with thermal radiation and chemical reaction, Commun. Nonlinear Sci. Numer. Simul. 15 (2010) 1813–1830.

5. K.-L. Hsiao, Heat and mass mixed convection for MHD viscoelastic fluid past a stretching sheet with Ohmic dissipation, Commun. Nonlinear Sci. Numer. Simul. 15 (2010) 1803–1812.

6. B.R. Rout, S.K. Parida, H.B. Pattanayak, Effect of radiation and chemical reaction on natural convective MHD flow through a porous

medium with double diffusion, J. Eng. Thermophys. 23 (2014) 53–65.

7. M. Modather, A.M. Rashad, A.J. Chamkha, An analytical study of MHD heat and mass transfer oscillatory flow of a micropolar fluid over a vertical permeable plate in a porous medium, Turkish J. Eng. Env. Sci. 33 (2009) 113.

8. M.A. El-Hakiem, MHD oscillatory flow on free convection radiation through a porous medium with constant suction velocity, J. Magn. Magn. Mater. 220 (2000) 271–276.

9. D. Pal, B. Talukdar, Buoyancy and chemical reaction effects on MHD mixed convection heat and mass transfer in a porous medium with thermal radiation and Ohmic heating, Commun. Nonlinear Sci. Numer. Simul. 15 (2010) 2878– 2893.

10. V. Sugunamma, N. Sandeep, P.M. Krishna, R. Bahunadam, Inclined magnetic field and chemical reaction effects on flow over a semi infinite vertical porous plate through porous medium, Commun. Appl. Sci. 1 (2013) 1–24.

11. A.K. Acharya, G.C. Dash, S.R. Mishra, Free convective fluctuating MHD flow through porous media past a vertical porous plate with variable temperature and heat source, Phys. Res. Int. 2014 (2014) article ID 587367.

12. K.V. Prasad, K. Vajravelu, A. Sujatha, Influence of internal heat generation/ absorption, thermal radiation, magnetic field, variable fluid property and viscous dissipation on heat transfer characteristics of a Maxwell fluid over a stretching sheet, J. Appl. Fluid Mech. 6 (2013) 249–256.

13. A.G. Vija, B. Rushi Kumar, B. Reddappa, K. Raveendra Babu, S.V.K. Varma, The effects of induced magnetic field and viscous dissipation on MHD mixed convective flow past a vertical plate in the presence of thermal radiation, Int. J. Appl. Eng. Res. 9 (18) (2014) 4533–4547.

14. K.-L. Hsiao, Heat and mass transfer for micropolar fluid with radiation effect past a nonlinearly stretching sheet, Heat Mass Transf. 46 (2010) 413– 419.

15. S. Shateyi, S.S. Motsa, P. Sibanda, The effects of thermal radiation, Hall currents, Soret, and Dufour on MHD flow by mixed convection over a vertical surface in porous media, Math. Probl. Eng. 2010 (2010) Article ID 627475- 20.

16. D. Pal, H. Mondal, Effects of Soret Dufour, chemical reaction and thermal radiation on MHD non-Darcy unsteady mixed convective heat and mass transfer over a stretching sheet, Commun. Nonlinear Sci. Numer. Simul. 16 (2011) 1942–1958.

17. P.M. Kishore, V. Rajesh, S. Vijayakumar Verma, The effects of thermal radiation and viscous dissipation on MHD heat and mass diffusion flow past an oscillating vertical plate embedded in a porous medium with variable surface conditions, Theor. Appl. Mech. 39 (2012) 99–125.

18. S. Karthikeyan, M. Bhuvaneswari, S. Rajan, S. Sivasankaran, Thermal radiation effects on MHD convective flow over a plate in a porous medium by perturbation technique, Appl. Math. Comp. Internl. 2 (1) (2013) 75–83.

19. M.S. Hossain, M.A. Samand, Heat and Mass Transfer of an MHD free convection flow along a stretching sheet with chemical reaction, radiation and heat generation in presence of magnetic field, Res. Math. Stat. 5 (1–2) (2013) 5–17.

20. K.-L. Hsiao, Nanofluid flow with multimedia physical features for conjugate mixed convection and radiation, Comput. Fluids 104 (20) (2014) 1–8.

21. M.A. El-Aziz, Mixed convection flow of a micropolar fluid from an unsteady stretching surface with viscous dissipation, J. Egypt. Math. Soc. 21 (2013) 385–394.

22. P.M. Kishore, D. Bhanumathi, S. Vijayakumar Verma, The influence of chemical reaction and viscous dissipation on unsteady MHD free convection flow past an exponentially accelerated vertical plate with variable surface conditions, Chem. Ind. Chem. Eng. Q. 19 (2) (2013) 181–193.

23. P.K. Singh, J. Singh, MHD flow with viscous dissipation and chemical reaction over a stretching porous plate in porous medium, Int. J. Eng. Res. Appl. 2 (2) (2012) 1556–1564.

24. A.C. Eringen, Theory of micropolar fluids, Math. Mech. 16 (1966) 1–18.

25. M.Z. Haque, M.M. Alam, M. Ferdows, A. Postelnicu, Micropolar fluid behaviors on steady MHD free convection and mass transfer flow with constant heat and mass fluxes, Joule heating and viscous dissipation, J. King Saud Univ. Eng. Sci. 24 (2012) 71–84.

26. M. Sudheer Babu, J. Girish Kumar, T. Sankar Reddy, Mass transfer effects on unsteady MHD convection flow of micropolar fluid past a vertical moving porous plate through porous medium with viscous dissipation, Int. J. Appl. Math. Mech. 9 (6) (2013) 48–67.

27. D. Gupta, L. Kumar, B. Singh, Finite element solution of unsteady mixed convection flow of micropolar fluid over a porous shrinking sheet, Scientificworldjournal 2014 (2014) article ID 362351.

CHAPTER 9

In silico modelling of mass transfer & absorption in the human gut

T.E. Moxon, O. Gouseti, S. Bakalis

Chemical Engineering, The University of Birmingham, Edgbaston, Birmingham, B15 2TT, UK

ABSTRACT

An *in silico* model has been developed to investigate the digestion and absorption of starch and glucose in the small intestine. The main question we are aiming to address is the relative effect of gastric empting time and luminal viscosity on the rate of glucose absorption. The results indicate that all factors have a significant effect on the amount of glucose absorbed. For low luminal viscosities (e.g. lower than 0.1 Pas) the rate of absorption is controlled by the gastric emptying time. For viscosities higher than 0.1 Pas a 10 fold increase in viscosity can result in a 4 fold decrease of glucose absorbed. Our model, with the simplifications used to develop it, indicate that for high viscosity luminal phases, gastric emptying rate is not the controlling mechanism for nutrient availability. Developing a mechanistic model could help elucidate the rate limiting steps that control the digestion process.

KEYWORDS

In silico modelling; Human gut; Glucose absorption; Mass transfer; Gastric emptying

1. INTRODUCTION

Understanding digestive processes is important in addressing diet related diseases, such as obesity, which are becoming a major problem all around the world. A World Health Organisation report in 2014 stated that 39% of adults were overweight and 13% were obese; also stating that the obesity rate was most prevalent in the Americas and least in the south-east Asian regions (WHO, 2014). Specifically in the UK around a quarter of adults were classified as obese as of 2014 (HSCIC, 2014); it has been estimated that obesity will cost the UK society £50billion per annum by 2050 (McPherson et al., 2007). In order to address some of the food related diseases and design healthier foods it is important to understand the behaviour of foods during digestion using *in silico*as well as *in vivo* and *in vitro* studies.

Modelling has been extensively used in a variety of systems e.g., pharmaceuticals (Peng and Cheung, 2009 and Stoll et al., 2000), biological systems such as the insulin-glucose system (Makroglou et al., 2006 and Pedersen and Cobelli, 2014). Simulation of biological processes allows for investigation into phenomena that are difficult to examine or study*in vivo* and *in vitro*. In this work we will be modelling digestion in the gut as a series of ideal reactors, a concept introduced in the late 1980's (Penry and Jumars, 1986 and Penry and Jumars, 1987), with wide applications in the area of pharmacokinetics (Ni et al., 1980, Peng and Cheung, 2009 and Stoll et al., 2000).

Mathematical models have been developed to investigate the digestion of foods using different approaches: A compartmental approach with a CSTR small intestine was used by Dalla Man et al. (2006) assuming that changes in gastric emptying rate have the largest effect on absorption (Dalla Man et al., 2006), this work showed good agreement with absorption from oral glucose tolerance tests. Bastianelli et al. (1996) simulated the movement and absorption of different nutrients simultaneously with a multiple compartmental approach (Bastianelli et al., 1996), which was able to predict nutrient absorption patterns and transit times. A model developed by Taghipoor et al. (2012) used a system of ODEs to simulate the movement and absorption from a food bolus within the intestine highlighting the effect dietary fibre has on slowing the bolus break down (Taghipoor et al., 2012 and Taghipoor et al., 2014).

Despite the fact that mathematical models provide insight into digestion; they typically use parameters that are obtained empirically, which limits their predictive capability.

1.1. Starch digestion

Starch is the largest source of carbohydrate in the human diet (Singh et al., 2010). In the small intestine, α-amylase will convert starch to oligosaccharides, and brush boarder enzymes (e.g., glucoamylase) will hydrolyse the oligosaccharides to glucose, which can then be absorbed. The conversion of oligosaccharide to glucose and absorption of glucose by sodium-dependent glucose cotransporter 1 (SGLT-1) proteins through the epithelium will be rapid and will not be rate limiting (Bastianelli et al., 1996, Lentle and Janssen, 2011 and Stumpel et al., 2001).

The kinetics of starch hydrolysis by α-amylase has been studied by a number of authors with the amylase substrate isolated from a variety of sources. Both bacterial and human α-amylase have been found to follow Michaelis–Menten kinetics (Ikram-Ul-Haq et al., 2010, komolprasert and Ofoli, 1991, Satomura et al., 1984 and Yankov et al., 1986), although it has also been reported that this is only followed for low substrate concentrations and at high concentrations a modified 1st order kinetics can be used (komolprasert and Ofoli, 1991). Inhibition of α-amylase by high D-glucose concentrations has been reported on some occasions (Steverson et al., 1984 and Yankov et al., 1986), which has been reported to have a large effect at concentration greater than 300 g/L (Yankov et al., 1986), though this is a high concentration that is unlikely to be encountered *in vivo.*

1.2. Gastric emptying

Gastric emptying rate is often considered to be the rate limiting step in the absorption of nutrients (Hellstrom et al., 2006 and Mourot et al., 1988). The delivery of gastric content to the duodenum is controlled by the pyloric sphincter (Hellstrom et al., 2006), whilst the stomach acts as a reservoir for consumed food, and mechanically and chemically breaks down the content (Kong and Singh, 2008).

Table 1 shows a selection of studies of the gastric emptying rate for different liquid solutions. Gastric emptying is quantified with a half-time (time for half the content to empty the stomach by volume) and calorific empting rate. These studies were selected as they have a comprehensive description of the physical properties of the fluids and the calorific content.

Table 1. Summary of gastric emptying data from literature, showing emptying rate for different liquid meals and the method of measurement, as well as comments to highlight the salient points of the studies.

Nutrient & thickener	Half-time [min]	Empty rate Kcal/min	Measurement method	Comments	Reference
500 mL 0.25 g/100 g LBG (μ_0: 0.01 Pa s)	17 ± 6	–	Echo-planar magnetic resonance imaging	• No significant variation of emptying time with changes in viscosity • Large changes in viscosity occurred in the stomach, pointing to the importance of gastric secretions • Over 40min, the viscosity of the 0.01 Pa s solution was reduced to 0.005 Pa s and that of the 11 Pa s solution was reduced to 0.3 Pa s.	(Marciani et al., 2000)
500 mL 0.5g/100 g LBG (μ_0: 0.1 Pa s)	18 ± 4	–			
500 mL 1.0g/100 g LBG (μ_0: 2 Pa s)	18 ± 7	–			
500 mL 1.5g/100 g LBG (μ_0: 11 Pa s)	19 ± 9	–			
500 mL–64 kcal, LV	32 ± 7	1	Echo-planar magnetic resonance imaging	• Slowing of gastric emptying observed with addition of nutrient for both HV & LV • HV low calorie solution emptied slower than LV, the effect was diminished for high calorie solutions, but still significant. • Antral volumes were higher with HV meals compared to LV meals	(Marciani et al., 2001)
500 mL–64 kcal, HV	46 ± 9	0.7			
500 mL–322.65 kcal, (63% lipid, 27% carbohydrate) LV	67 ± 9	2.4			
500 mL–322.65 kcal, (63% lipid, 27% carbohydrate) HV	79 ± 6	2.0			
600 mL–96 kcal Glucose LV	17 ± 1	2.8	Double sampling gastric aspiration technique	• Increase in emptying time (4–8 fold) with increased solution energy content (4 fold) • Show longer emptying for lower viscosity equicarbohydrate solutions, contrary to other authors.	(Vist and Maughan, 1995)
600 mL–96 kcal glucose HV	14 ± 1	3.4			
600 mL–451 kcal glucose LV	130 ± 18	1.7			
600 mL–451 kcal glucose HV	64 ± 8	3.5			
600 mL- 60 kcal glucose solution	9.4 ± 1.2	3.2	Double sampling gastric aspiration technique	• Linear relationship between the caloric density and calorific empty rate was observed • Main factor in the emptying rate is the calorific density	(Calbet and MacLean, 1997)
600 mL- 132 kcal PPH	16.3 ± 5.4	4.05			
600 mL–138 kcal WPH	17.2 ± 6.1	4.01			
600 mL–396 kcal MP	26.4 ± 10	7.5			
300 mL- 400 kcal glucose	107	1.9	Scintigraphy	• Solutions with high calories have longer emptying times • Solutions used varied in both volume and calorific content, hence making it difficult to identify the most important factor	(Phillips et al., 1991)
450 mL–200 kcal glucose	66	1.5			
500 ml-500 kcal (mixed) LV	72.1 ± 19.5	3.5	Ultra-sonography	• with higher viscosity solutions having slightly longer emptying times • Results here show large variability (~20–25%) • Calories are from mixed sources not just glucose	(Yu et al., 2014)
500 ml-500 kcal (mixed)	85.5 ± 16.5	2.9			
400 ml- 400 kcal (mixed) LV	257.9 ± 31.8	0.8	^{13}C breath sampling with continuous IR spectrometry	• Overall emptying faster for HV • Initial empty rate faster for LV • author linker this to inhibition due to nutrient sensing in the duodenum	(Shimoyama et al., 2007)
400 ml- 400 kcal (mixed) HV	195.1 ± 16.3	1.0			
400 ml-Water	99.4 ± 2.8	–			
240 kcal Solid/liquid meal	77 ± 6	1.56	Echo-planar magnetic resonance imaging	• Looked at effect of blended (soup) vs. Solid meal with water drink • Longer emptying for soup, linked by author to sieving mechanism whereby low nutrient liquid phase is able to empty separately from the high nutrient solid phase • The soup has homogenous nutrient composition and the emptying will stimulate the nutrient feedback mechanism, slowing the emptying rate.	(Marciani et al., 2012)
240 kcal Soup	92 ± 7	1.3			

LBG e locust bean gum, PPH- Pea peptide hydrolysate solution, WPH- Whey peptide hydrolysate solution, MP- Milk protein solution, LV e low viscosity, HV e High viscosity, 1 g glucose ¼ 4 kcal.

Table 2. Parameter values used in the model with references.

Parameter	Value	Reference
Surface area increase due to folds, villi & microvilli (f)	12	(Ganong, 2005; Stoll et al., 2000; Lentle and Janssen, 2011)
Mean velocity	1.7×10^{-4} m/s	(Stoll et al., 2000)
Length of small intestine	2.85m	(Stoll et al., 2000)
Radius of small intestine	1.8 cm	(Stoll et al., 2000)
Radius of glucose molecule (r_0)	0.38 nm	(Schultz and Solomon, 1961)
Simulation time	10800 s	
Initial glucose/starch mass	50 g	
Viscosity	0.001–10 Pa s	
Emptying half time	2min – 2h	
V_{max}	1-25 mM/min	(Fonseca, 2011; Satomura et al., 1984)
K_m	9 mM	(Fonseca, 2011)

In Table 1 the measurement methods can be separated in 3 groups: breath sampling, aspiration, and imaging (e.g. MRI/Scintigraphy/Sonography). The most common method for measuring gastric emptying rates in a medical setting is Scintigraphy, where meals are labelled with ^{99m}Tc, and distributions of these radioisomers are taken using gamma cameras (Punkkinen et al., 2006). Punkkinen et al. (2006) compared this to the ^{13}C breath test, where a meal is labelled with ^{13}C and breath samples are taken and the ratio of ^{13}C–^{12}C can be used to calculate the volume remaining in the stomach. The group found that the ^{13}C breath test gives significantly longer emptying half-time than Scintigraphy and that there was no correlation between the half-lives of the two methods (Punkkinen et al., 2006). This could explain why the results by Shimoyama et al. (2007)have longer emptying rates when compared to the rest of the table (also shown in Fig. 1).

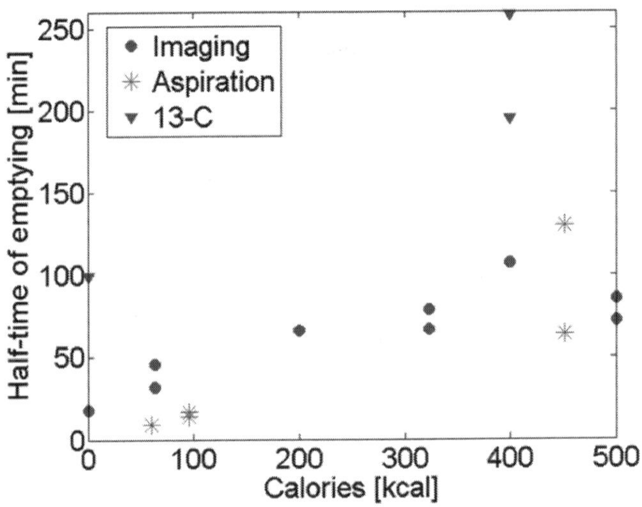

Figure 1. plot of half-time of emptying against calories for meals in Table 1, different colours represent different methods of measurements, showing that increasing the calorific content of a meal leads to a longer half time of emptying.

Scintigraphy has also shown 70% slower emptying rates than double sampling aspiration, where a dye is added to a meal and samples are taken directly from the stomach via catheter and emptying inferred (Beckers et al., 1992), although this is not evident from the data presented in Table 1. Good agreement in measured emptying rates with MRI (Feinle et al., 1999 and Schwizer et al., 1992) and ultrasonography (Hveem et al., 1996) are also shown in literature.

Fig. 1 shows a plot of half-time of emptying against the calorific content of the meal for different measurement methods. As one can see the resulting emptying times depend on the method of measurement. As previously explained the ^{13}C method results in a significantly higher estimation of gastric emptying time; this results in a large uncertainty on parameters used in models as a large variety of sources have to be considered typically each employing a different method.

As can be seen in Fig. 1, an increase in calorific content results in an increase of gastric emptying time, but the scarcity of the data points do not allow us to conclude upon the nature of the relationship. This could be explained from observations widely reported in literature of a feedback mechanism from the small intestine (controlled by nutrient sensors) that is thought to be the main controlling mechanism of gastric emptying rate (Brener et al., 1983, Calbet and MacLean, 1997, McHugh, 1983 and Shimoyama et al., 2007).

Whilst in Table 1 there is a clear trend with emptying rate and the calorific content, the link between the emptying rate and viscosity or volume of meal consumed is not clear. Prior to the initiation of this feedback mechanism, there is an initial rapid emptying rate, which is independent of the nutrient content of the meal. Some researchers suggest that this rate will be controlled by the volume of fluid in the stomach (Brener et al., 1983 and Moran et al., 1999), while others point to the effect of viscosity (Marciani et al., 2001 and Shimoyama et al., 2007), with more viscous meals causing greater distension of the antral region relative to the proximal; and also resulting in a great volume of gastric secretions (Marciani et al., 2001). However, contradictory results on the effect of viscosity on gastric emptying have been reported, as seen in Table 1. The effect of gastric secretions could play a key role in determining gastric viscosity (see for example (Marciani et al., 2000)).

The last two results in Table 1 indicate the difference in emptying between two meals of the same constitution, one in solid/liquid form and one as a soup. There is a difference between how solids and how liquids will empty from the stomach, with solids requiring a reduction in particle size, to around 1–2 mm, before they can empty (Hellstrom et al., 2006). The current work will focus on the ingestion of liquid meals and the gastric processes will not be considered, other than the emptying to the small intestine.

1.3. Modelling of absorption in the small intestine

Within the intestinal lumen the chyme (mixture of consumed food and secretions from the digestive system) will be propelled aborally and via peristaltic contractions, which may also provide mixing of the nutrients (Janssen et al., 2007). Segmentation contractions will mix the chyme with no movement axially along the intestine (Ganong, 2005). The flow of nutrients along the digestive tract has been studied by numerous authors using computation fluid dynamics (CFD). Studies have been carried out to look at the mixing effects in the stomach (Ferrua and Singh, 2010, Ferrua and Singh, 2011 and Kuzo et al., 2010), the flow at the gastroduodenal junction (Dillard et al., 2007), and the flow in the intestine (Love et al., 2013, Nadeem et al., 2012, Riahi and Roy, 2011, Tripathi, 2011 and Tripathi et al., 2011). These studies indicate that flow dynamics will affect the movement of nutrients to the luminal wall; this mass transfer can be an important parameter in nutrient bioaccessibility (whether the nutrients are in a form which can be absorbed).

In silico (computer simulated) studies of absorption in the small intestine have been carried out for drug and foods using different methodologies. In pharmacokinetics, two main types of models have been used: non-compartmental and compartmental. Non-compartmental models are generally developed by fitting a mathematical expression to*in vivo* data, hence the fitted parameters will be accurate only for the system analysed and will not offer any predictive capability. In compartmental models, the system is divided in to compartments each representing a different physiological process; each with different mathematical expressions. A well formulated model should offer a certain amount of predictive capability (Peng and Cheung, 2009).

In literature the small intestine has been modelled as a single compartment (Dalla Man et al., 2006 and Di Muria et al., 2010), as multiple compartments (Bastianelli et al., 1996 and Yu et al., 1996) or as a plug flow reactor (PFR) (Logan et al., 2002 and Stoll et al., 2000).

The model developed by Dalla Man et al. (2006) attempted to simulate an oral glucose tolerance test and a test meal. The intestine was modelled as a continuous stirred tank reactor (CSTR) with an input from the emptying of the stomach which is a function of the mass of glucose in the stomach. The model required the fitting of 6 parameters and had overall a good agreement with experiments ($p < 0.005$). However, a more physiologically relevant control for gastric emptying is the sensing of nutrients in the duodenum, followed by the relevant feedback response (Brener et al., 1983 and Calbet and MacLean, 1997). In addition, there is no consideration of food properties, which are likely to affect absorption (Gouseti et al., 2014 and Tharakan et al., 2010), this will be one of the focus of the models developed in this work.

Yu et al. (1996) compared different compartmental and plug flow models, concluding that the flow profile in the small intestine can be characterised with both a multi-compartmental model and a plug flow model, but that a single-compartmental model, as used by Dalla Man et al. (2006) was inadequate at describing the flow profile.

A multi-compartmental model was developed by Bastianelli et al. (1996) for a meal containing a variety of nutrients; the model used 4 compartments described by a series of ordinary differential equations.

Although successful in describing digestion of a complex meal, the model does not consider the effect of one component on another (e.g. the effect of fibre on bioaccessibility of nutrients), the effect of nutrients on stomach emptying, nor the spatial location or movement along each compartment (Bastianelli et al., 1996). Another method, used by Stoll et al. (2000), was a plug flow model for the absorption of drugs in the small intestine. This model included the effect of increased surface area due to the folds and projections on the surface of the small intestine, and the effect of eddy rolls, to give good agreement with the absorption and degradation within the systemic circulation system (Stoll et al., 2000). The model does not include any gastric disintegration or emptying effects, which might have significant implications on the absorption of nutrients.

The movement and degradation of a bolus in the small intestine was investigated byTaghipoor et al. (2012). The model considered the effect of non-degradable and soluble nutrients; it was further developed to look at the effect of dietary fibre (Taghipoor et al., 2014). This will have an effect on the viscosity and water holding capacity of the bolus, and on the absorption of nutrients (Taghipoor et al., 2014). This highlights the importance of bioaccessibility during digestion.

An all-in-one model would allow full representation of physiological conditions, however this will be at the expense of simplicity, increase the difficulty of implementation and requiring a large number of parameters, with doubtful advantages over *in vivo* tests. The different factors need indeed to be considered separately, and with their relative importance (Calbet and MacLean, 1997, Hellstrom et al., 2006 and McHugh, 1983).

In this paper a mechanistic approach to the modelling of mass transport and absorption from the small intestine is attempted; focusing on the effect of the delivery of the nutrients to the small intestine from the stomach, the mass transfer (as a function of viscosity of the chyme) within the lumen of the small intestine and the hydrolysis of starch prior to absorption. This study will focus on the use of a plug flow reactor small intestine, assuming a laminar flow and constant mean velocity. In reality the regime of flow will depend on the nature of the chyme; lower viscosity solutions exhibiting turbulent flow, and more viscous solutions displaying laminar flow with large regular vortices (Janssen et al., 2007 and Lentle and Janssen, 2008), as a result of wall contractions and curvature of the small intestine. Hence the assumption of laminar flow will likely underestimate the mass transfer of nutrients at particular viscosities and also affect the residence time distribution (Janssen et al., 2007), but the comparison of this parameter to the gastric emptying rate, and hydrolysis rate should still be revealing.

This paper will present three models of increasing complexity. In the first model, mass transfer of nutrients (exemplified by glucose) within the small intestine and through the intestinal wall will be linked with the viscosity of chyme. The effect of gastric emptying on glucose absorption will then be considered in the second model. The third model will include starch hydrolysis, assuming the reaction to follow Michaelis–Menten kinetics.

2. DEVELOPMENT OF MODELS

The following models (models 1–3) have been developed to investigate different factors that could influence the absorption of nutrients: bioaccessibility within the small intestine, gastric emptying rate and hydrolysis rate.

The models assume that the stomach and small intestine can be described by a series of reactors, specifically a continuous stirred tank reactor (CSTR) for the stomach, which will act as a reservoir and control the emptying of contents only, and a plug flow reactor (PFR) for the small intestine (Fig. 2).

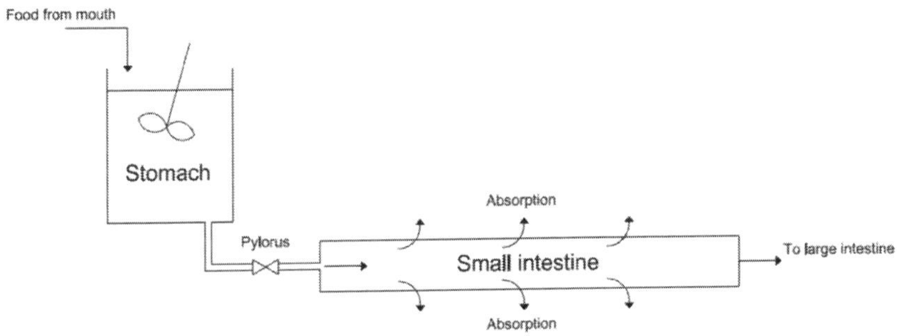

Figure 2. diagram showing layout of CSTR stomach and PFR small intestine.

The models will be developed with increasing complexity, the first looking at the effect of mass transfer within the lumen on absorption of nutrients; the next will include the mass transfer and gastric emptying rate and how these both affect the absorption rate; the final model will look at, mass transfer rate, gastric emptying rate and starch hydrolysis, and how all 3 effect the absorption of nutrients.

2.1. Model 1: glucose absorption

This first model aims to investigate the effect of mass transfer on glucose absorption in the small intestine; this was modelled as a 1D advection-reaction equation (Logan et al., 2002):

$$\frac{\partial G(z,t)}{\partial t} = -\overline{u}\frac{\partial G(z,t)}{\partial z} - \frac{2f}{r_m}KG(z,t) \tag{1}$$

Change in glucose mass with time = movement along SI due to advection = Absorption of glucose

Initial conditions:

$$G(z,0) = \begin{cases} G_0 & for\ l_0; \\ 0 & otherwise; \end{cases} \quad (2)$$

Boundary conditions:

$$\left.\frac{\partial G}{\partial z}\right|_{z=0} = \left.\frac{\partial G}{\partial z}\right|_{z=L} = 0 \quad (3)$$

Here $G(z,t)$ is the glucose concentration at time *t*, and distance along the intestine *z*, and $\bar{u}$, is the mean velocity along the length of the intestine. The last term is the absorption of glucose, where K, is the mass transfer coefficient, *2/rm* is the ratio of surface area to volume for a cylinder and *f* represents the increase in absorptive surface area due to the folds of the intestinal wall. It is assumed that the volume input is a bolus and enters the small intestine at position *l0* from the entrance, which is equal the radius of the bolus of entering liquid.

The overall mass transfer coefficient, *K,* will depend on the mass transfer within the lumen, through the epithelium layer and into the blood (Tharakan et al., 2010). As we are mainly interested in bioaccessibility we will simplify the phenomena and will ignore the effect of transport through the epithelium and blood assuming they are rapid and not rate limiting (Bastianelli et al., 1996, Stumpel et al., 2001 and Wang et al., 2010). Therefore mass transfer coefficient, *K* is calculated from the relationship between Sherwood (Sh=Kd/D), Reynolds ($Re = \rho \bar{u} d/\mu$) and Schmidt (Sc=μ/ρD) numbers, where *d* is the mean intestinal diameter, *L* is the length of the intestine, *D* is the diffusivity, *ρ* is the density, and *μ* is the viscosity. The flow is in the laminar regime (for a water like solution, Re ~ 3), and the following empirical relationship is used (Carbonell, 1975).

$$Sh = 1.62\, Re^{1/3} Sc^{1/3} \left(\frac{d}{L}\right)^{1/3} \quad (4)$$

Rearranging in terms of *K* gives:

$$K = 1.62\left(\frac{\overline{u}D^2}{Ld}\right)^{1/3} \tag{5}$$

The diffusivity is calculated from the Einstein-Stokes equation, which will depend on the viscosity of the system:

$$D = \frac{K_B T}{6\pi\mu r_0} \tag{6}$$

Here *KB* is the Boltzmann constant, *T* is the absolute temperature (310 K), and *r0* is the radius of the diffusing molecule.

Therefore the only parameter that we can control and manipulate will be the viscosity of the food, and from this we can manipulate the mass transfer rate. The mass transfer rate will be inversely proportional to the viscosity to the power of 2/3 i.e.:

$$K \propto \frac{1}{\mu^{2/3}} \tag{7}$$

Looking at the effect of protrusions on the surface of the intestine, it can be approximated that the villi increase the surface area by around 10 times relative to a cylinder and the microvilli by around 20 times (Ganong, 2005 and Stoll et al., 2000). But only around 2% of the surface will be involved with the absorption of glucose, due to the fast speed of the absorption (Lentle and Janssen, 2011), giving an increased surface area of 4 times that of a cylinder. Including the effect of increased surface area from the presence of plicae circulares estimated at 3 times (Ganong, 2005 and Stoll et al., 2000), this will give a value of *f* as 12. Values for parameters used in the models are shown in Table 2.

The results, typically of glucose absorbed can also be described in terms of calories where 1 g of glucose is 4 kcal.

The equation can also be made dimensionless, glucose was expressed as a fraction of the inlet concentration ($G' = G/G0$), time was divided by the residence time to give τ (=t u/L), and the distance along the intestine was divided by the length to give ξ (=z/L):

$$\frac{\partial G'(\xi,\tau)}{\partial \tau} = -\frac{\partial G'(\xi,\tau)}{\partial \xi} - \tau_{transfer} G'(\xi,\tau) \tag{8}$$

Where,

$$\tau_{transfer} = \frac{2fK}{r_m}\frac{L}{\overline{u}} \tag{9}$$

This yields the dimensionless number $\tau_{transfer}$ which is the characteristic time of mass transfer, i.e. the mass transfer rate ($2fK/rm$) multiplied by the mean residence time of passage through the small intestine ($L/\ \overline{u}$).

2.2. Model 2: stomach emptying and intestinal absorption of glucose

The gastric empting rate is thought to be the controlling mechanism in absorption of nutrients (Mourot et al., 1988 and Hellstrom et al., 2006); for this reason a model was built to estimate the overall effect of the gastric emptying and mass transfer of glucose in the small intestine.

This model will treat the stomach as a reservoir for delivery of nutrients to the intestine only and will not consider its effect on the structure (chemical or physical) of the food. Gastric emptying is modelled as exponential decay, i.e. a liquid solution with no lag phase (Calbet and MacLean, 1997 and Hellstrom et al., 2006), as this model shows a good approximation of the emptying for liquid only meals. The model for the intestine will be the same as for model 1 but with an input from the gastric emptying.

The glucose mass in the stomach was represented by G_S:

$$\frac{\partial G_S}{\partial t} = -\gamma G_S \tag{10}$$

$$G_S|_{t=0} = G_{S0} \tag{11}$$

The model for the small intestine will take the following form:

$$\frac{\partial G(z,t)}{\partial t} = \begin{cases} \gamma G_s - \overline{u}\frac{\partial G(z,t)}{\partial z} - \frac{2f}{r_m}KG(z,t), & \text{if } z = l_0 \\ -\overline{u}\frac{\partial G(z,t)}{\partial z} - \frac{2f}{r_m}KG(z,t), & \text{otherwise} \end{cases} \tag{12}$$

$$G(z,0)=0; \tag{13}$$

And the boundary conditions:

$$\left.\frac{\partial G}{\partial z}\right|_{z=0} = \left.\frac{\partial G}{\partial z}\right|_{z=L} = 0 \tag{14}$$

Where G_{S0} is the initial input of glucose to the stomach (50 g) and, γ, is the decay constant, which can be expressed as the half-time of emptying, which is a common parameter used to describe the emptying of liquids from the stomach:

$$t_{1/2} = \frac{\ln(2)}{\gamma} \tag{15}$$

The model can also be made dimensionless in the same way as the advection-reaction equation to give:

$$\frac{\partial G_S'(\tau)}{\partial \tau} = -\tau_{empting} G_S'(t) \tag{16}$$

Where,

$$\tau_{emptying} = \gamma \frac{L}{\overline{u}} \tag{17}$$

Here $\tau_{emptying}$ is the characteristic time of gastric emptying and represents the rate of gastric emptying against the residence time in the small intestine. The half emptying times were varied between 10min and 3 h, which is within the range of typical values (seen in Table 1). The characteristic time of emptying was varied between 0.5 and 100, and the characteristic time of mass transfer was varied between 0.1 and 100, to see the effect on the fraction of glucose absorbed after the time is equivalent to the residence time.

2.3. Model 3: starch hydrolysis

In this work we will assume the starch remains intact until it reaches the small intestine, at which point hydrolysis of the starch, following Michaelis–Menten kinetics, will occur and a mass balance on starch and glucose in the small intestine has been carried out. In the small intestine model á-amylase will be in excess, and the ability to hydrolyse will be limited by the bioaccessibility of enzyme to starch, hence will be limited not by amount of enzyme but by the properties of the chyme (Ballance et al., 2013 and Englyst and Englyst, 2005). The effect of salivary α-amylase is not included as the focus of the study is the hydrolysis in the intestine and hence the input is starch only into the stomach at t = 0. The model will therefore take a form similar to model 2, with an extra component of starch:

$$\frac{\partial S_s}{\partial t} = -\gamma S_s \tag{18}$$

$$\frac{\partial S(z,t)}{\partial t} = \begin{cases} \gamma S_s - \overline{u}\dfrac{\partial S(z,t)}{\partial z} - \dfrac{V_{max} S(z,t)}{K_m + S(z,t)}, & \mathit{if}\ z = l_0 \\ -\overline{u}\dfrac{\partial S(z,t)}{\partial z} - \dfrac{V_{max} S(z,t)}{K_m + S(z,t)}, & \mathit{otherwise} \end{cases} \tag{19}$$

Change in starch mass with time = movement along SI due to advection e starch converted to glucose

$$\frac{\partial G(z,t)}{\partial t} = -\bar{u}\frac{\partial G(z,t)}{\partial z} + \frac{V_{max}S(z,t)}{K_m + S(z,t)} - \frac{2f}{r_m}KG(z,t) \tag{20}$$

Change in glucose mass with time = movement along SI due to advection + generation of glucose - absorbed glucose Initial conditions and boundary conditions are the same as model 2, with input of starch.

These equations can be made dimensionless:

$$\frac{\partial S'(\xi,\tau)}{\partial \tau} = -\frac{\partial S'(\xi,\tau)}{\partial \xi} - \frac{\tau_R S'(\xi,\tau)}{K_{mII} + S'(\xi,\tau)} \tag{21}$$

$$\frac{\partial G'(\xi,\tau)}{\partial \tau} = -\frac{\partial G'(\xi,\tau)}{\partial \xi} + \frac{\tau_R S'(\xi,\tau)}{K_{mII} + S'(\xi,\tau)} - \tau_{transfer}G'(\xi,\tau) \tag{22}$$

This yields two more dimensionless numbers as well as $\tau_{transfer}$:

$$\tau_R = \frac{L}{\bar{u}}\frac{V_{max}}{Gs0} \tag{23}$$

$$K_{mII} = \frac{K_m}{Gs0} \tag{24}$$

τ_R is the characteristic time of reaction, which is the residence time in the small intestine multiplied by the maximum reaction rate scaled with the initial input of starch. K_{mII} is the Michaelis constant normalised with the initial input of starch.

The characteristic time of reaction will be varied from 1 to 25 as the characteristic emptying was varied between 0.5 and 100 and characteristic time of mass transfer was varied between 0.1 and 100 to see the effect on fractional absorption of glucose after the time is equal to the residence time.

2.4. Simulations

The equations for each model were simulated in gPROMS (v.3.7.1); the partial differentials were solved using backward finite difference method. All models were simulated over a period of 3 h, similarly to what is used for glycaemic index measurements (Brouns et al., 2005 and Wolever et al., 1991). Graphs were produced using MATLAB (R2014a).

3. RESULTS & DISCUSSION

3.1. Model 1

The first model investigated the mass transfer of glucose in the small intestine, from an initial input at t = 0 of 50 g of glucose solution at different viscosities (20 simulations for viscosities ranging 0.001 Pa s and 10 Pa s). The lower viscosity corresponds to viscosity of water, while the higher viscosity would be relevant to honey. Fig. 3(a) shows the amount of absorbed glucose against time. The initial rate of absorption decreases as the majority of glucose is absorbed, the effect being more pronounced at low viscosities. The results in terms of the rate of calories absorbed can be seen in Fig. 3(b). For low viscosities one can see an initial high rate of absorption as the luminal glucose is absorbed. From Fig. 3(a), it appears that by around 1 h, about half of the 50 g input of glucose has been absorbed and by around 3 h about 80% has been absorbed. This will result to a lower amount of glucose in the lumen, and a lower absorption gradient, which explains the decrease in the absorption rate for the low viscosity solution. When higher viscosity solutions were used the absorption rate stays almost constant over the 3 h simulated. This is due to the low mass transfer rate resulting in only a small percentage of the luminal glucose being absorbed. In Fig. 3(c) the effect of viscosity on absorbed glucose after 3 h is shown. Fig. 3(c) indicates that at low viscosities (1 mPa s) glucose is absorbed to a high extent (~80% of input), and as the viscosity increases the amount of glucose absorbed decreases. For viscosity higher than 0.1 Pa s, the total absorbed glucose is less than 10% of the input and does not significantly reduce with viscosity.Fig. 3(d) is a dimensionless representation of Fig. 3(c), i.e. glucose absorbed versus viscosity/rate of mass transfer. The curve has a sigmoidal shape, showing a rapid increase between $\tau_{transfer}$ values of 0.1 and 3, corresponding to viscosity values of about 0.2 Pa s and 1 mPa s.

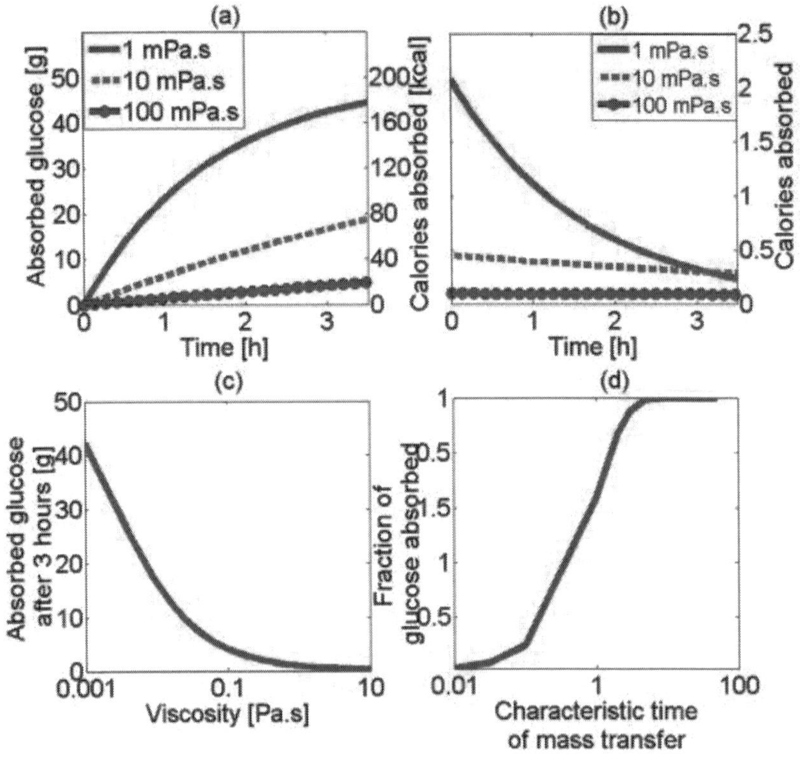

Figure 3. (a) the absorption curves for glucose solutions at different viscosities; (b) graph showing the total absorbed glucose after a 3 h period for solutions of different viscosities (log scale); (c) the fraction of glucose absorbed for the non-dimensionilised model against the characteristic mass transfer coefficient(log scale); (d) the rate at which calories are absorbed at different viscosities.

At low values of $\tau_{transfer}$, i.e., rate of absorption slow compared to the residence time, little absorption of glucose occurs; as $\tau_{transfer}$ increases an increase in absorbed glucose is observed. As the value of $\tau_{transfer}$ reaches 1, i.e., the rate of mass transfer is similar to the rate of advection along the length of the intestine, a plateau occurs with total absorption of the fed glucose.

These results indicate that the mass transfer coefficient within the lumen (determined by luminal viscosity) may have a large effect on the absorption of nutrients in the small intestine especially when mass transfer is significantly limiting the rate of absorption typically at viscosity values greater than 0.1 Pa s. Similar relationships have been seen*in vivo*, for example, Ellis et al. (1995) showed a non-linear relationship between zero shear viscosity of the chyme (measured in the jejunum) and absorption of nutrients in pigs, also indicating an inverse linear relationship between absorption over a 4 h period and concentration of guar gum in the meal (Ellis et al., 1995). Takahashi et al. (2009)showed how the disappearance of glucose in the small intestine of rats is inversely proportional to the viscosity (Takahashi et al., 2009), in this work 3 different viscosities were used and so there is not enough evidence to extrapolate these models. Fitting these results to a power–law curve the absorption was found inversely proportional to the viscosity to the power of around 0.45; this is lower than the relationship suggested in the present model (see equation (7)). This difference could be due to the added motility of a functioning gut, when digesting materials of high viscosity which is not included in this model. Due to secretions in the stomach and intestine the viscosity is unlikely to be constant with time, which is another limitation of the current model, but the trends here are consistent with those reported in the literature. Leclere et al. (1994) took a different view and speculated that the observed changes in blood glucose etc. from different viscosity meals are mainly due to the effect of the viscosity on stomach emptying rather than any mass transfer resistance within the small intestine (Leclere et al., 1994), which the current results show.

Model 2 will be used to compare the emptying rate and mass transfer rate to test this hypothesis.

Overall in this work the parameters used were obtained from literature and the results were within the range of order of magnitude seen with *in vivo* data from literature see for example Dalla Man et al. (2006). Validation of similar *in silico* digestion models can be challenging as availability rather than postprandial glucose data would be required. As part of on going work we are aiming to make best use of existing *in vivo* data in the literature to validate our models. *In vitro* studies though demonstrate some agreement with the results presented here. Tharakan et al. (2010) showed a decrease in absorption with viscosity, pointing to an increase in diffusion resistance or decreased mixing efficacy as an explanation (Tharakan et al., 2010). A 50% decrease in absorption was seen when the guar gum was added at 0.5% compared to a starch mix with no guar gum. Gouseti et al. (2014) showed similar results for the absorption of glucose *in vitro* from model solutions for a range of food hydrocolloids. Others show similar trends (Sasaki and Kohyama, 2012, Singh et al., 2010 and Slaughter et al., 2002), whilst speculating that the viscosity modifiers may have additional effects on the digestion process such as encapsulation of starch molecules, thus reducing bioaccessibility (Sasaki and Kohyama, 2012), or direct inhibition of digestive enzymes (seen with Guar Galactomannan) (Slaughter et al., 2002). The results here seem to agree with many of the experimental observations but the present model does not include any mixing effects, e.g., via segmentation or peristalsis. These are likely to increase the mass transfer rate and hence increase the absorption rate (Gouseti et al., 2014 and Tharakan et al., 2010), and subsequently this could reduce the effect of viscosity upon the absorption rate. In future work the effect of mixing could be included in the mass transfer coefficient by investigating how segmentation/peristaltic mixing will affect the empirical relationship between Sherwood, Reynolds, and Schmidt numbers.

3.2. Model 2

The second model expanded on model 1 by including the effect of gastric emptying, on glucose absorption. The input at t = 0 is into the stomach, and not into the intestine; the stomach then feeds the intestine. Fig. 4(a) shows the estimated emptying for glucose solutions for 3 different emptying half-times (15 min, 30 min and 1 h shown). Increasing the half-time results in a slower empting rate, as expected. Fig. 4(b) shows the associated glucose absorption in the small intestine. The total absorbed has a sigmoidal shape. For times smaller than 15mins the rate of absorption is low as expected from the small amount of glucose in the lumen (more than 50% glucose still been in the stomach). This is equivalent to an induction time. This is followed by an almost linear increase as more glucose enters the intestine and is available to be absorbed. The rate of absorption decreases after the majority of luminal glucose is absorbed. As the half-time of emptying increases, the induction time decreases and the rate of absorption decreases. Fig. 4(c) is a contour plot showing absorption of glucose versus emptying times (characteristic time of emptying) and viscosities (characteristic time of mass transfer). This plot can be separated into 4 regions: (1) the bottom right shows an area where the emptying rate is limiting, and greater characteristic time of emptying will result in greater absorption of glucose and vice versa; (2) the bottom left region shows an area where both emptying rate and mass transfer rate will be rate limiting; (3) the top left shows the area where mass transfer rate will be limiting only; and (4) top right area shows the area where near maximum absorption is reached (these regions are shown more clearly in Fig. 5).

The characteristic mass transfer time was varied from 0.1 to 100. The values of 0.1 and 3.4 corresponding to viscosities of 0.2, 10^{-3} Pa s, respectively; in this range of viscosities we would expect that mass transfer can be the rate limiting step. The higher values of characteristic mass transfer rates are in regime of effective and rapid mixing, i.e. one where mass transfer values are very large, e.g. $K = 1 \times 10^{-6}$ m/s The characteristic time of emptying was varied between 0.5 and 100; where the value of 3.2 and 100 resulted in half emptying time of 1 h and 2 min respectively. The lower value of 0.5 was included to investigate what happens for slow emptying and fast intestinal transit, corresponding to a 2 h emptying half-time and 1.5 h (Read et al., 1986) intestinal residence time.

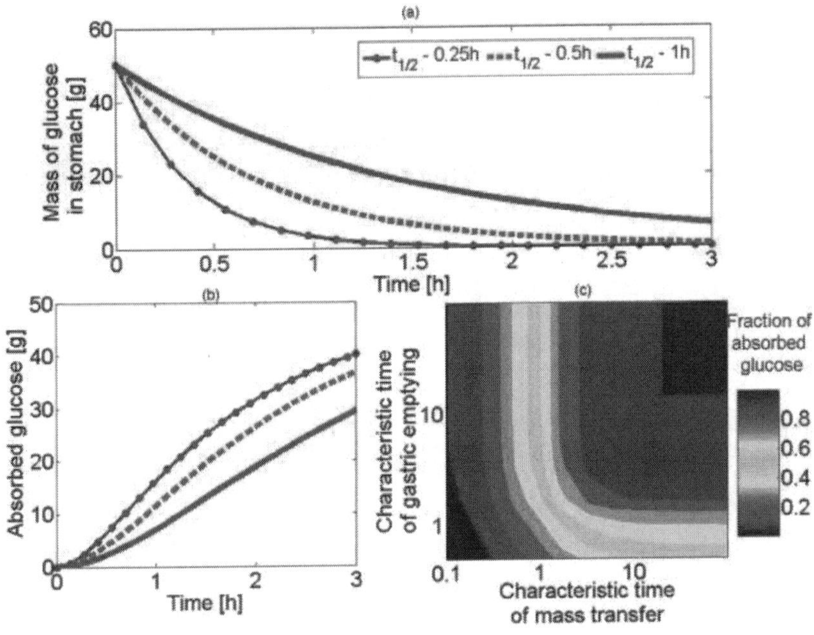

Figure 4. (a) mass of glucose in stomach over time with different half-time's of emptying and viscosity of 1 mPa s, (b) the absorbed glucose against time for 3 different gastric emptying half-time's, (c) contour plot of the characteristic mass transfer, against the characteristic emptying time on log–log scale, colour representing the fraction of glucose absorbed.

In this work we will incorporate published research to understand the effect of gastric emptying (Brener et al., 1983, Calbet and MacLean, 1997, Marciani et al., 2000, Marciani et al., 2001 and Shimoyama et al., 2007), using exponential decay to model the stomach emptying (Brener et al., 1983, Calbet and MacLean, 1997 and Hellstrom et al., 2006). For reasons of simplicity the effect of secretions is not included in this model (Marciani et al., 2000 and Marciani et al., 2001)

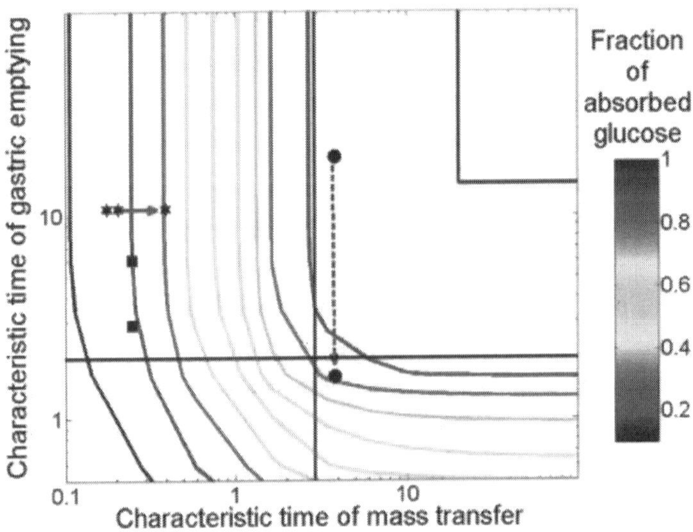

Figure 5. contour plot from Fig. 4(c) with plots from literature (*) (Marciani et al., 2000), (■) (Marciani et al., 2001), (●) from the model.

The two square data points (Marciani et al., 2001) represent low and high nutrient meals with similar viscosities. One would expect that they will have similar fractional absorption of glucose, even with a change in gastric emptying rate, as absorption is controlled from mass transfer. The majority of the data in Table 1 for liquid meals will appear into the upper left region of Fig. 5, indicating that the total absorption after around 3 h should be mass transfer limited.

The final set of points connected by a blue dotted arrow indicates how solutions, with similar initial viscosities (1 mPa s), would be affected by increasing the gastric emptying half-time from 10min to 2 h. In this case the system will see little effects in the total glucose absorption until it crosses the black horizontal line (around a half-time of 1 h). A further increase in the gastric emptying half-time will result in a shift to a gastric emptying limited region, which will cause a reduction in the fraction of glucose absorbed. Overall for relatively high viscosity food systems it appears that the fraction of glucose absorbed after 3 h is not controlled from gastric emptying rate, as with half-times of less than 90min (a characteristic time of emptying around 2), the system will be in the upper left region ofFig. 5, limited by the mass transfer coefficient.

3.3. Model 3

Model 3 incorporates the effect of starch hydrolysis to produce glucose on Model 2. InFig. 6(a) absorbed glucose is plotted against time for different rates of hydrolysis, V_{max}. One can see that the increasing V_{max} results in an increase in glucose absorption and decrease of lag phase (the initial slow absorption region). This is expected as the faster the starch is hydrolysed to glucose the faster glucose can be absorbed. However, increasing chyme viscosity will also affect the bioavailability of starch for reaction, or enzyme kinetics. In Fig. 6(b) contour plots of glucose absorbed for different characteristic reaction rates against characteristic emptying and mass transfer are shown.

The planes show similarities to Fig. 4(c), where the plot of characteristic emptying and transfer rate showed four regions. In Fig. 6(b), at low characteristic reaction rates there is little change in absorption as either emptying or mass transfer are changed, i.e., very little of the starch is hydrolysed to glucose, but increasing the reaction rate moves the system away from being reaction limited and the other parameters have a greater effect on glucose absorption, at around a characteristic reaction rate of 25, the starch is hydrolysed very quickly and behave similar to the Fig. 4(c), where the input is purely glucose. The two slices in the middle, V_{max} = 7.1 mmol/min (Satomura et al., 1984) and V_{max} = 14.1 mmol/min (Fonseca, 2011), show results for reaction kinetics taken from literature, and it can be seen that the reaction rate can be limiting if these rates are seen*in vivo*.

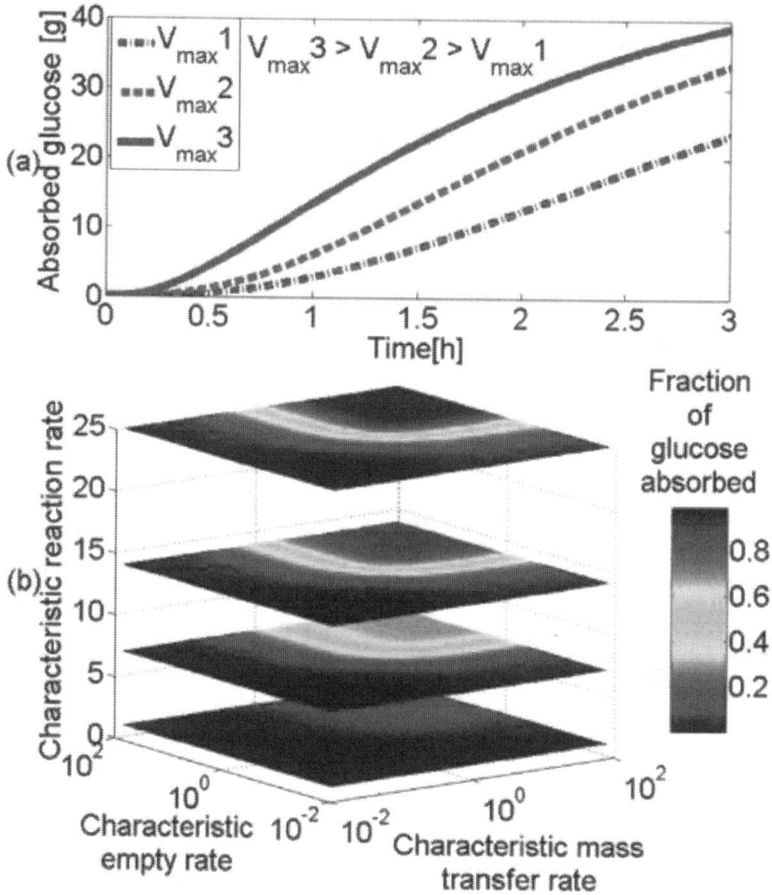

Figure 6. (a) absorption of glucose with time for systems with different starch hydrolysis rates (gastric emptying half-time 20min, viscosity = 1 mPa s, V_{max} = 4, 9 and 16 mmol/min), (b) contour plot showing the effect of gastric emptying rate, mass transfer rate and reaction rate for hydrolysis on absorption of glucose.

Each of these parameters is currently independent of the others, but in reality they are likely coupled. Changes in viscosity are likely to affect the emptying and mass transfer of nutrients as previously stated, as well as mass transfer of the enzymes. In addition, It is also important to consider other effects of food ingredients e.g., nutrient encapsulation by thickeners or direct enzyme inhibition by additives (Sasaki and Kohyama, 2012 and Slaughter et al., 2002).

4. CONCLUSION

Mathematical models to describe *in vivo* digestion were developed and used to examine the relative effect of gastric emptying, mass transfer and reaction rate limitations in the small intestine. Within the assumptions of the models the results indicate that for gastric emptying half-times less than 1 h the viscosity/mass transfer rate is the limiting factor for the amount of glucose absorbed after 3 h. If the emptying half-time is greater than 1 h, both the gastric emptying and mass transfer rates can influence the absorption depending on the viscosity. If the mass transfer rate is faster than 1×10^{-7} m/s (i.e. luminal viscosity of 1 mPa s), the amount absorbed in 3 h is not limited by the mass transfer, and only by the gastric emptying rate. Starch hydrolysis reaction rates, when both the mass transfer and gastric emptying are fast and not limiting, can have a pronounced effect. The reaction kinetics for starch hydrolysis from literature showed around 25% difference in absorption when used in the model. Further development of the models is required to understand some of the controlling mechanisms as well as comparison with *in vivo* data to obtain confidence in the validity of the results.

ACKNOWLEDGEMENTS

The authors would like to acknowledge BBSRC, DRINC, grant reference BB/I006079/1for providing financial support for this work.

REFERENCES

1. Ballance, S., et al., 2013. Evaluation of gastric processing and duodenal digestion of starch in six cereal meals on the associated glycaemic response using an adult fasted dynamic gastric model. Eur. J. Nutr. 52, 799e812.

2. Bastianelli, D., Sauvant, D., Rerat, A., 1996. Mathematical modeling of digestion and nutrient absoprtion in pigs. J. Animal Sci. 74, 1873e1887.

3. Beckers, E.J., Leiper, J.B., Davidson, J., 1992. Comparison of aaspiration and scintigraphic techniques for the measurement of gastric emptying rates of liquids in humans. Gut 33, 115e117.

4. Brener, W., Hendrix, T.R., McHugh, P.R., 1983. Regulation of the gastric emptying of glucose. Gastroenterology me 85, 76e82.

5. Brouns, F., et al., 2005. Glycaemic index methodology. Nutr. Reasearch Rev. 18, 145e171.

6. Calbet, J.A.L., MacLean, D.A., 1997. Role of caloric content on gastric emptying in humans. J. Physiology 498 (2), 553e559. Carbonell, R.G., 1975. Mass transfer coefficients in colied tubes. Biotechnol. Bioeng. 17, 1383e1385.

7. Dalla Man, C., Camilleri, M., Cobelli, C., 2006. A system model of oral glucose Absorption:Validation on Gold Standard data. IEEE Trans. Biomed. Eng. 53 (12), 2472e2478.

8. Di Muria, M., Lamberti, G., Titomanlio, G., 2010. Physiologically based pharmacokinetics: a simple, all purpose model. Industrial Eng. Chem. Res. 49, 2969e2978.

9. Dillard, S., Krishnan, S., Udaykumar, H.S., 2007. Mechanics of flow and mixing at antroduodenal junction. World J. Gastroenterol. 13 (9), 1365e1371.

10. Ellis, P.R., Roberts, F.G., Low, A.G., Morgan, L.M., 1995. The effect of high-molecularweight guar gum on net apparent glucose absorption and net apparent insulin and gastric inhibitory polypeptide production in the growing pig: relationship to rheological changes in jejunal digesta. Br. J. Nutr. 74, 539e556.

11. Englyst, K.N., Englyst, H.N., 2005. Carbohydrate bioavailability. Br. J. Nutr. 94, 1e11. Feinle, C., et al., 1999. Scintigraphic validation of a magnetic resonance imaging method to study gastric emptying of a solid meal in humans. Gut 44, 106e111.

12. Ferrua, M.J., Singh, R.P., 2010. Modeling the fluid dynamics in a human stomach to gain insight of food digestion. J. Food Sci. 75 (7), R151eR162.

13. Ferrua, M.J., Singh, R.P., 2011. Understanding the fluid dynamics of gastric digestion using computational modeling. Procedia Food Sci. 1, 1465e1472.

14. Fonseca, M.R.J., 2011. An Engineering Understanding of the Small Intestine (thesis), s.l.. University of Birmingham.

15. Ganong, W., 2005. Gastrointestinal physiology. s.l. In: ganong's Review of Medical Physiology. The McGraw-Hill Companies, pp. 429e486.

16. Gouseti, O., et al., 2014. Hydrocolloids in human digestion: dynamic in-vitro assessment of the effect of food formulation on mass transfer. Food Hydrocoll. 42, 378e385.

17. Hellstrom, P.M., Gryback, P., Jacobsson, H., 2006. The physiology of gastric emptying. Best Pract. Res. Clin. Anaesthesiol. 20 (3), 397e407.

18. HSCIC, 2014. Statistics on obesity, physical activity and diet: England. Health and Social Care Information Centre.

19. Hveem, K., Jones, K.L., Chatterton, B.E., Horowitz, M., 1996. Scintigraphic measurment of gastric emptying and ultrasonographic assessment of antral area: relation to appetite. Gut 38, 816e821.

20. Ikram-Ul-Haq, Javed, M.M., Hameed, U., Adnan, F., 2010. Kinetics and thermodynamic studies of alpha amylase from bacillus licheniformis mutant. Pak. J. Bot. 42 (5), 3507e3516.

21. Janssen, P.W.M., et al., 2007. Characterisation of flow and mixing regimes within the ileum of brushtail possum using residence time distribution analysis with simultaneous spatio-temporal mapping. J. Physiology 582, 1239e1248.

22. komolprasert, V., Ofoli, R.Y., 1991. Starch hydrolysis kinetics of bacillus licheniformis alpha-amylase. J. Chem. Technol. Biotechnol. 51, 209e223.

23. Kong, F., Singh, R.P., 2008. Disintegration of solid foods in human stomach. J. food Sci. 73 (5), R67eR80.

24. Kuzo, H., et al., 2010. Analysis of flow phenomena in gastric contents induced by human gastric peristalsis using CFD. Food Biophys. 5, 330e336.

25. Leclere, C.J., et al., 1994. Role of viscous guar gums in lowering the glycemic response after a solid meal. Am. J. Clin. Nutr. 59, 914e921.

26. Lentle, R.G., Janssen, P.W., 2008. Physical characteristics of digesta and their influence on flow and mixing in the mammalian intestine: a review. J. Comp. Physiol. B 178, 673e690.

27. Lentle, R.G., Janssen, P.W., 2011. The Physical Processes of Digestion. s.l.. Springer.

28. Logan, D.J., Joern, A., Wolesensky, W., 2002. Location, time and temperature dependence of digestion in simple animal tracts. J. Theor. Biol. 216, 5e18.

29. Love, R.J., et al., 2013. An expanded finite element model of the intestinal mixing of digesta. Food Dig. 4, 26e35.

30. Makroglou, A., Li, J., Kuang, Y., 2006. Mathematical models and software tools for the glucose-insulin regulatory system and diabetes: an overview. Appl. Numer. Math. 56, 559e573.

31. Marciani, L., et al., 2000. Gastric response to increased meal viscosity assessed by echo-planar magnetic resonance imaging in humans. J. Nutr. 30 (1), 122e127.

32. Marciani, L., et al., 2001. Effect of meal viscosity and nutrients on satiety, intragastric dilution, and emptying assessed by MRI. Am. J. Physiol. Gastrointest. Liver Physiol. 280, G1227eG1233.

33. Marciani, L., et al., 2012. Preventing gastric sieving by blending a solid/water meal enhances satiation in healthy humans. J. Nutr. 142 (7), 1253e1258.

34. McHugh, P.R., 1983. The control of gastric emptying. J. Aut. Nerv. Syst. 9, 221e231. McPherson, K., Marsh, T., Brown, M., 2007. Modelling future trends in obesity and the impact on health.

35. Moran, T.H., Wirth, J.B., Schwartz, G.J., McHugh, P.R., 1999. Interactions between gastric volume and duodenal nutrients in the control of liquid gastric emptying. Am. J. Physiology 276 (4), R997eR1002.

36. Mourot, J., et al., 1988. Relationship between the rate of gastric emotying and glucose and insulin responses to strachy foods in young healthy adults. Am. J. Clin. Nutr. 48, 1035e1040.

37. Nadeem, S., Ashiq, S., Ali, M., 2012. Williamson fluid model for the peristaltic flow of chyme in small intestine. Math. Probl. Eng. 1e18. Ni, P.F., et al., 1980. Theoretical model studies of intestinal drug absorption V. Nonsteady-state fluid flow and absorption. Int. J. Pharm. 5, 33e47.

38. Pedersen, M.G., Cobelli, C., 2014. Insulin modelling. s.l.. In: Modelling Methodology for Physiology and Medicine. Elsevier, pp. 333e353.

39. Peng, H., Cheung, B., 2009. A Review on Pharmacokinetic Modeling and the Effects of Environmental Stressors on Pharmacokinetics for Operational Medicine. Defence R&D Canada, Toronto. Penry, D.L., Jumars, P.A., 1986. Chemical reactor analysis and optimal digestion. BioScience 36 (5), 310e315.

40. Penry, D.L., Jumars, P.A., 1987. Modeling animal guts as chemical reactors. Am. Nat. 129 (1), 69e96.

41. Phillips, W.T., Schwartz, J.G., Blumhardt, R., McMahan, C.A., 1991. Linear gastric emptying of hyperosmolar glucose solutions. J. Nucl. Med. 32, 377e381.

42. Punkkinen, J., et al., 2006. Measuring gastric emptying: comparison of C-13- octanoic acid breath test and scintigraphy. Dig. Dis. And Sci. 51 (2), 262e267.

43. Read, N.W., et al., 1986. Simulatmeous measurement of gastric emptying, small bowel residence and colonic filling of a solid meal by use of the gamma camera. Gut 27, 300e308.

44. Riahi, D.N., Roy, R., 2011. Mathematical modeling of peristaltic flow of chyme in small intestine. Appl. Appl. Math. 6 (2), 428e444.

45. Sasaki, T., Kohyama, K., 2012. Influence of non-starch polysaccharides on the in vitro digestibility and viscosity of starch suspensions. Food Chem. 133, 1420e1426.

46. Satomura, S., et al., 1984. Kinetics of human pancreatic and salivary alpha-amylase with carboxymethylamyloses as substrates. Clin. Chim. Acta 138, 21e29. Schultz, S.G., Solomon, A.K., 1961. Determination of the effective hydrodynamic radii of small molecules by viscometry. J. General Physiology 44, 1189e1199.

47. Schwizer, W., Maecke, H., Fried, M., 1992. Measurement of gastric emptying by magnetic resonance imaging in humans. Gastroenterology 103 (2), 369e379.

48. Shimoyama, Y., et al., 2007. High-viscosity liquid meal accelerates gastric emptying. Neurogastroenterol. Motil. 19, 879e886.

49. Singh, J., Dartois, A., Kaur, L., 2010. Starch digestibility in food matrix: a review. Trends Food Sci. Technol. 21, 168e180.

50. Slaughter, S.L., Ellis, P.R., Jackson, E.C., Butterworth, P.J., 2002. the effect of guar galactomannan and water availability during hydrothermal processing on the hydrolysis of starch catalysed by pancreatic alpha-amylase. Biochim. Biophys. Acta 1571, 55e63.

51. Steverson, E.M., Korus, R.A., Admassu, W., Heimsch, R.C., 1984. Kinetics of the amylase system of Saccharomycopsis fibuliger. Enzyme Microb. Technol. 6, 549e554.

52. Stoll, B.R., et al., 2000. A theory of molecular absorption from the small intestine. Chem. Eng. Sci. 55, 473e489.

53. Stumpel, F., Burcelin, R., Jungermann, K., Thorens, B., 2001. Normal kinetics of intestinal glucose absorption in the absence of GLUT2: evidence for a transport pathway requiring glucose phosphorylation and transfer into the endoplasmic reticulum. Proc. Natl. Acad. Sci. 98 (20), 11330e11335.

54. Taghipoor, M., et al., 2012. Mathematical modeling of transport and degradation of feedstuffs in the small intestine. J. Theor. Biol. 294, 114e121.

55. Taghipoor, M., et al., 2014. Digestion modelling in the small intestine: Impact of dietary fiber. Math. Biosci. 258, 101e112.

56. Takahashi, T., et al., 2009. Hydrolyzed guar gum decreases postprondial blood glucose and glucose absorption in the rat small intestine. Nutr. Res. 29, 419e425.

57. Tharakan, A., Norton, I.T., Fryer, P.J., Bakalis, S., 2010. Mass transport and nutrient absorption in a simulated model of small intestine. J. Food Sci. 75 (6), E339eE346.

58. Tripathi, D., 2011. A mathematical model for the peristaltic flow of chyme movement in small intestine. Math. Biosci. 233, 90e97.

59. Tripathi, D., Pandey, S.K., Das, S., 2011. Peristaltic transport of a generalized burgers' fluid: Application to the movement of chyme in small intestine. Acta Astronaut. 69, 30e38.

60. Vist, G.E., Maughan, R.J., 1995. The effect of osmolality and carbohydrate content on the rate of gastric emptying of liquids in man. J. Physiol. 486 (2), 523e531.

61. Wang, Y., et al., 2010. A multiscale lattice boltzmann model of macro- to micro-scale transport, with applications to gut function. Philos. Trans. R. Soc. 368, 2863e2880.

62. WHO, 2014. Global Status Report on Noncommunicable Diseases 2014. WHO, Geneva. Wolever, T.M., Jenkins, D.J., Jenkins, A.L., Josse, R.G., 1991. The glycemic index: methodology and clinical implications. Am. J. Clin. Nutrit. 54, 846e854.

63. Yankov, D., Dobreva, E., Beschkov, V., Emanuilova, E., 1986. Study of optimum conditions and kinetics of starch hydrolysis by means of thermostable alphaamylase. Emzyme Microb. Technol. 8, 665e667.

64. Yu, K., et al., 2014. The impact of soluble dietary fibre on gastric emptying, postprandial blood glucose and insulin in patients with type 2 diabetes. Asia Pac. J. Clin. Nutr. 23 (2), 210e218.

65. Yu, L.X., Crison, J.R., Amidon, G.L., 1996. Compartmental transit and dispersion model analysis of small intestinal transit flow in humans. Int. J. Pharm. 140, 111e118.

Index